ETHYLENE

Keystone to the Petrochemical Industry

CHEMICAL INDUSTRIES

A Series of Reference Books and Text Books

Consulting Editor

HEINZ HEINEMANN

Heinz Heinemann, Inc.,
Berkeley, California

Volume 1: Fluid Catalytic Cracking with Zeolite Catalysts, *Paul B. Venuto and E. Thomas Habib, Jr.*

Volume 2: Ethylene: Keystone to the Petrochemical Industry, *Ludwig Kniel, Olaf Winter, and Karl Stork*

Additional Volumes in Preparation

ETHYLENE

Keystone to the Petrochemical Industry

Ludwig Kniel · Olaf Winter · Karl Stork
C–E Lummus
Combustion Engineering, Inc.
Bloomfield, New Jersey

MARCEL DEKKER, INC. New York and Basel

Library of Congress Cataloging in Publication Data

Kniel, Ludwig, [Date]
Ethylene, keystone to the petrochemical industry.

(Chemical industries; v.2)
Includes index.
1. Ethylene. I. Winter, Olaf, [Date] II. Stork, Karl
[Date] III. Title. IV. Series.
TP248.E72K54 665′.733 79-22090
ISBN: 0-8247-6914-7

Marcel Dekker, Inc.
270 Madison Avenue, New York, New York 10016

Current printing (last digit):
10 9 8 7 6 5 4 3 2 1

Printed in the United States of America

Preface

Since the late 1930s a new industry has come into being in the United States that has gradually taken root and expanded in all of the industrialized countries over the world. Before that time ethylene, the simplest olefinic compound, had been recovered in relatively small quantities from coke oven gas as a by-product along with ammonia in operations to produce metallurgical coke for the steel industry. The raw materials base for these operations was coal.

The discovery that olefins could be produced in large yields from hydrocarbons contained in natural gas at reasonable cost and that methane, its principal constituent, was a ready source for synthesis gas, from which (concurrent with then available knowledge in catalysis), methanol, ammonia and a host of chemical intermediates could be obtained, initiated a gradual but consequently revolutionary reorientation in the chemical industry. The growth and expansion that followed in the wake of the change-over to a hydrocarbon basis from coal represents a chapter that remains unique in the annals of industrial development. Hand in hand with it went the development of new technologies aimed at the substitution of olefins, particularly of ethylene, for the more reactive acetylene whose raw materials basis was limestone and coal.

Today this industry has come to be known as the petrochemicals industry. From a production measured in thousands of tons annually during the mid-thirties, the United States production of ethylene alone has, due to the successful introduction of new products, soared to about 10.2 million tons annually in 1976. The nonsocialist world, exclusive of the United States, in 1976 produced 16.2 million tons of ethylene. Single train olefins plants of capacities corresponding to well in excess of 500,000 tons per annum of ethylene are no longer a rarity, and the number of useful products and intermediates originating from these operations extends into the thousands.

Occupying a prominent place in the petrochemical industry are the large oil and gas companies who historically are the producers and purveyors of the feedstock materials needed for the production of olefins. The chemical industry, at one time largely coal-based, had to reorient itself to seek liaison in various ways with the producers of oil and gas to assure for themselves an adequate share of the available supplies of these hydrocarbons.

The effective development and expansion of the petrochemical industry into a major manufacturing sector, in a time span of little more than a generation, could only take place through the availability and devoted efforts of researchers and operating staff in the production companies, of engineers and technicians in the engineering and construction industry and their supplier companies, and of cadres of skilled field personnel that erected the plants to achieve these ever-expanding production goals.

The book which is herewith presented to a technically oriented audience attempts to provide an overview of the important phases of the manufacture of olefins from naturally occurring hydrocarbons. It contains chapters on the chemistry involved, reaction kinetics, and the fundamental unit operations in the production, handling, storage, and transportation of ethylene. Also discussed are the raw materials situation and other related subjects. The contributors to this volume are members of the Lummus Technical Center at Bloomfield, New Jersey. They have drawn on the collective experience of the company in the design of some 93 olefins plants, a major share of the world's production capacity. In addition, there is available a vast reservoir of published work in this field, and the relatively large reference section at the end of the volume attests in part to their attempt to make the presentation as representative as possible.

To have the text reflect the latest status of the art, inquiries to test the accuracy of quoted information were sent to a number of companies active in either the design or operation of olefins plants. Chapter 8 on "The Pyrolysis of Hydrocarbons" was reviewed by several of these organizations. Due acknowledgment for their cooperative attitude in confirming data quoted in their behalf is made to the firms of Pullman-Kellogg, Linde A.G. Werksgruppe Muenchen, and Mitsubishi Petrochemicals Co.

As far as is known, no text of comparable scope has appeared in the technical literature. It is hoped that the present volume fills this gap and that it will be of assistance to professionals and students alike in broadening their understanding of the many facets of this technically challenging and still evolving technology.

Bloomfield, N. J.
April 1978

William Tucker
Senior Vice President
C-E Lummus
Combustion Engineering, Inc.

Contributors

Olefins technology in all its ramifications involves the application of scientific principles and artful elements from several disciplines, mainly chemical and mechanical engineering, chemistry, and metallurgy.

The manuscript of this volume was produced as a cooperative venture by individuals who, in the service of the Lummus Company, have spent a considerable part of their professional career to help advance this technological frontier.

Name	*Chapter*
Mr. Michael Abrahams, Manager, Economics and Cost Engineering	*10*
Dr. Meng-Teck Eng, Senior Engineer/Economist	*10*
Mr. Ludwig Kniel, Consultant, Process Technology	*5, 7, 8, 10*
Dr. L. Michael Shipman, Group Leader, Thermophysical Properties	*2*
Mr. Karl Stork, Manager, Process Technology	*1, 6, 7, 8, 9*
Dr. Chung-Hu Tsai, Senior Chemical Engineer	*6, 8*
Mrs. Karen Wagner, Secretary to the Group of Contributors	—
Dr. Olaf Winter, Principal Engineer, Process Planning	*3, 4, 6*

Due acknowledgment is made of helpful suggestions by internal reviewers of the manuscript, Messrs. S. M. Solomon, J. M. O'Donnell, C. H. Viens, K. Muenz, and L. Krawitz, of the Lummus Technical Center, Bloomfield, New Jersey.

Contents

Note: Units of measurement of The International System of Units (SI System) or acceptable equivalents are used throughout this volume. All tons (t) are metric tons (1000 kg).

1. Introduction

Ethylene, in the Geneva nomenclature "ethene" ($H_2C{=}CH_2$), is the lightest olefinic hydrocarbon. At ambient conditions it is a colorless, flammable gas of the same density as nitrogen with a slightly sweet odor. Ethylene does not freely occur in nature and yet represents the organic chemical consumed in the greatest quantity worldwide. It is the basic chemical raw material for a large variety of industrial products, either directly as in the production of polyethylene, or after further reaction with other chemicals as in the production of polyvinyl chloride, polystyrene, and polyesters.

The importance of ethylene for the chemical industry is founded on its reactivity due to the double bond in its molecular structure. Although less reactive than acetylene but simpler, safer, and less costly to produce and further convert, ethylene has largely replaced acetylene as the basic building block for an entire branch of industrial organic chemistry.

The growth of ethylene production over the last 25 years has been spectacular, based on the abundant availability of inexpensive light hydrocarbon feedstocks such as ethane and propane derived from natural gas and naphtha derived from crude oil. Until the early 1970s, technological advances and the erection of increasingly larger and more economical production facilities have resulted in a steady decline of ethylene production costs, spurring the consumption of ethylene in the production of its derivatives. The recent large cost increases of both raw materials and process equipment have resulted in an abrupt reversal of this trend.

Despite many economic uncertainties surrounding the petrochemical industry, ethylene production and consumption are expected to grow at a faster rate than the world economy due to the continuing replacement of natural and inorganic materials with organic synthetics and the further development of radically new synthetic materials.

2. Physical Properties

The thermodynamic and physical properties of ethylene (Table 1) have been and continue to be the subject of intensive investigations. A comprehensive treatise on the thermodynamic data of ethylene was published by Angus et al. [1], and a reliable set of thermodynamic charts and tables was presented by Starling et al. [2, 3].

The tabulation of Angus covers the temperature range of 0 to 150°C and

TABLE 1 Thermophysical Properties of Ethylene

Molecular weight	28.0536
Triple point:	
Temperature	−169.19°C
Pressure	0.11 kPa
Latent heat of fusion	3,350 J/mol
Normal boiling point:	
Temperature	−103.71°C
Latent heat of vaporization	13,540 J/mol
Density of the liquid	20.27 mol/dm^3
Specific heat of the liquid	67.4 J mol^{-1} K^{-1}
Viscosity of the liquid	1.61×10^{-4} Pa · s
Surface tension of the liquid	0.0164 N/m
Specific heat of the ideal gas (at 25°C)	42.84 J mol^{-1} K^{-1}
Critical point:	
Temperature	9.2°C
Pressure	5.042 MPa
Density	7.635 mol/dm^3
Compressibility factor	0.2813
Gross heat of combustion of the gas at 25°C	1.411 MJ/mol
Limits of flammability at atmospheric pressure and 25°C:	
Lower limit in air	2.7 mol%
Upper limit in air	36.0 mol%
Auto ignition temperature in air at atmospheric pressure	490°C

pressures from 0 to 300 MPa. Starling's tables cover temperatures from −150 to 260°C and pressures from 0 to 55 MPa.

Benzler and v. Koch, using correlations of the reduced properties at supercritical conditions and certain experimental data available to them, have calculated and published in chart form the enthalpy and entropy values of ethylene in the range up to 400°C and 1000 MPa [204].

Following the consensus of American experts from government, industry, and the universities on the further need of investigations on ethylene [4], a joint industry–government research project on the "thermodynamic properties of ethylene" was established. This project is administered by the office of Standard Reference Data of the National Bureau of Standards and will cover temperatures from −170 to 250°C and pressures to 300 MPa.

The first results of this program have been published [5–9], and completion of the entire project is scheduled for late 1979.

The available literature data on the viscosity and thermal conductivity of ethylene have been presented in chart form by Bonscher et al. [10].

Figures 1–6 show the more important temperature- and/or pressure-dependent characteristics of ethylene: Fig. 1, vapor pressure; Figs. 2a, 2b, and 2c, density; Figs. 3a, 3b, and 4, enthalpy; Fig. 5, viscosity; Fig. 6, thermal conductivity.

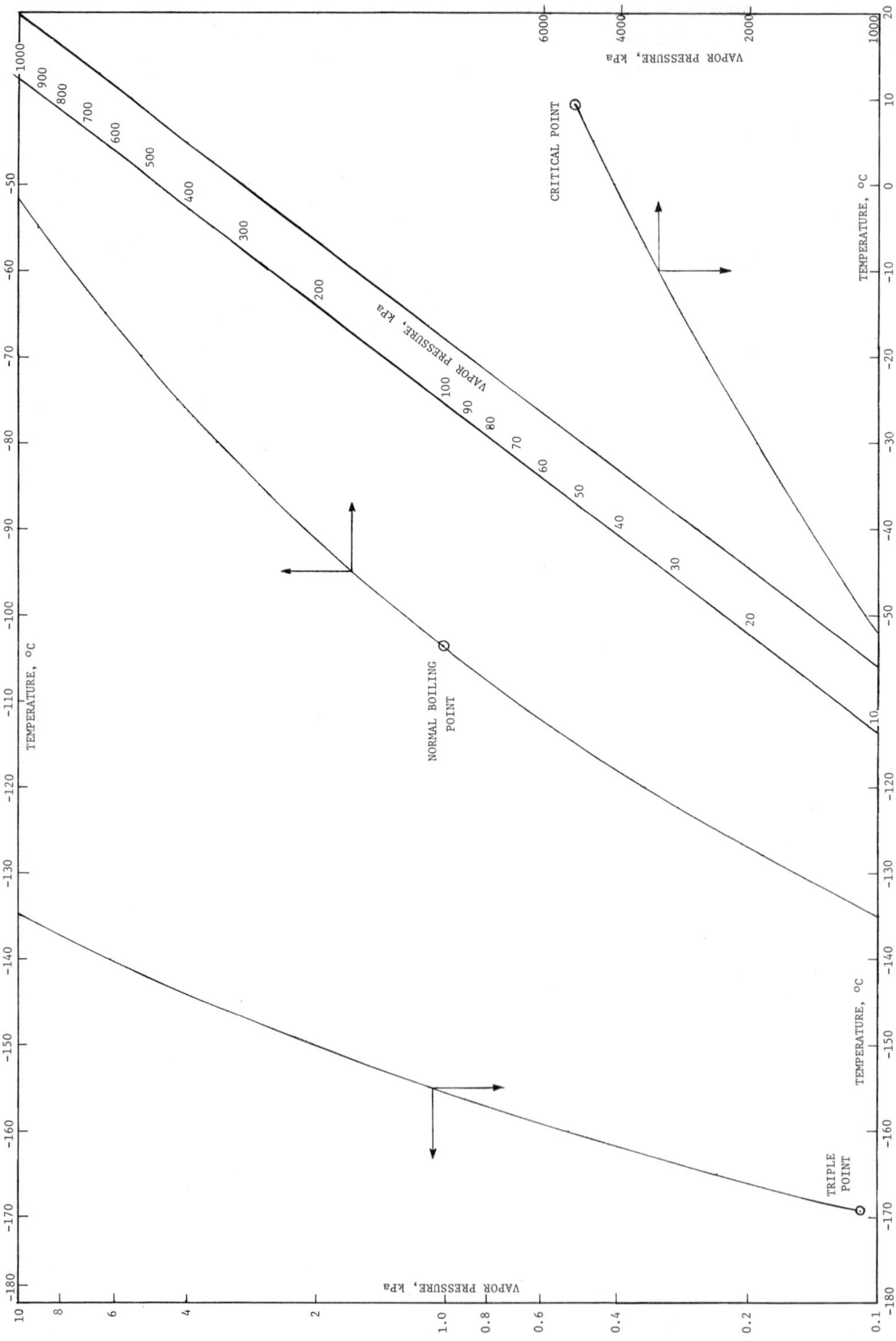

FIG. 1. Vapor pressure of ethylene.

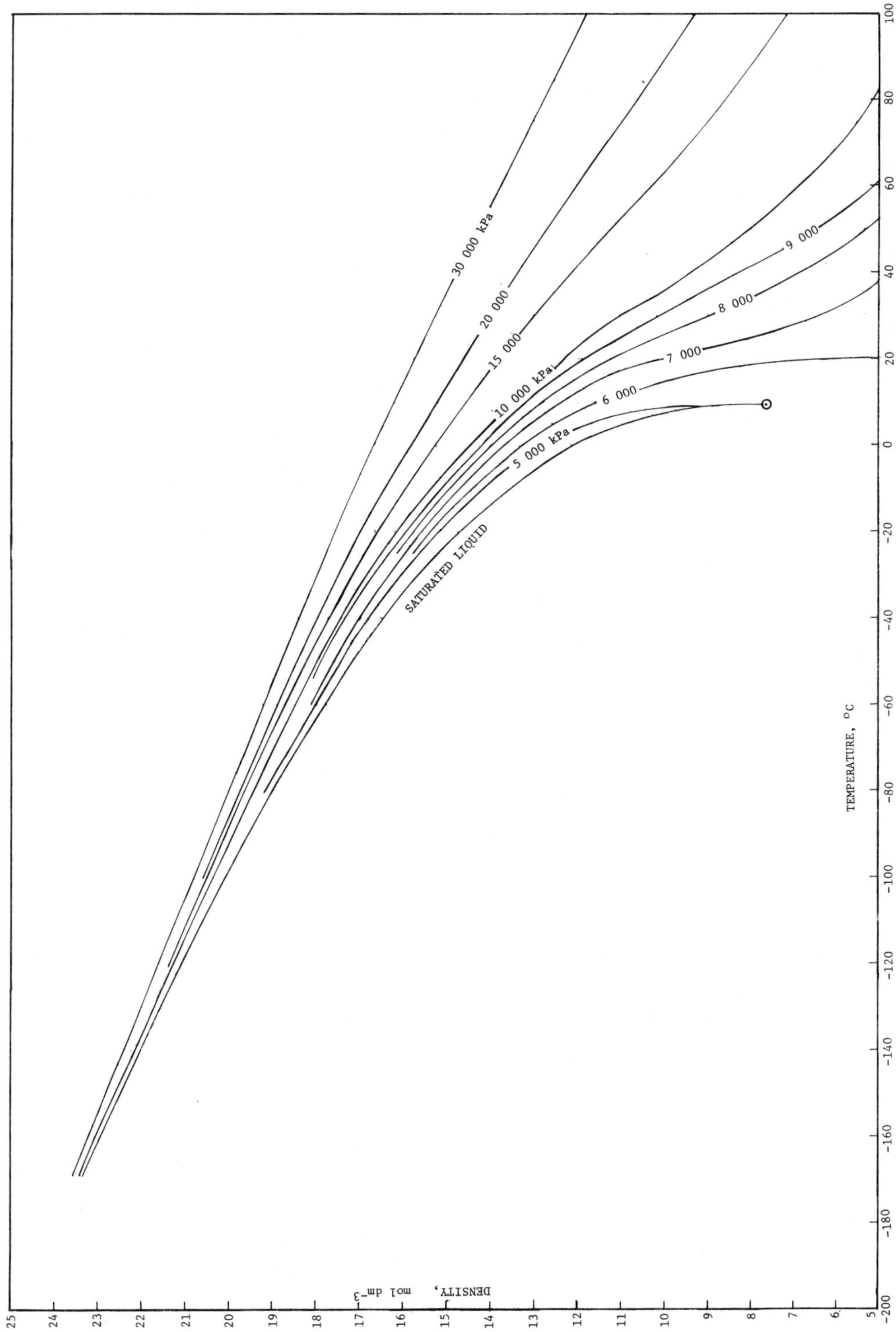

FIG. 2a. Density of ethylene.

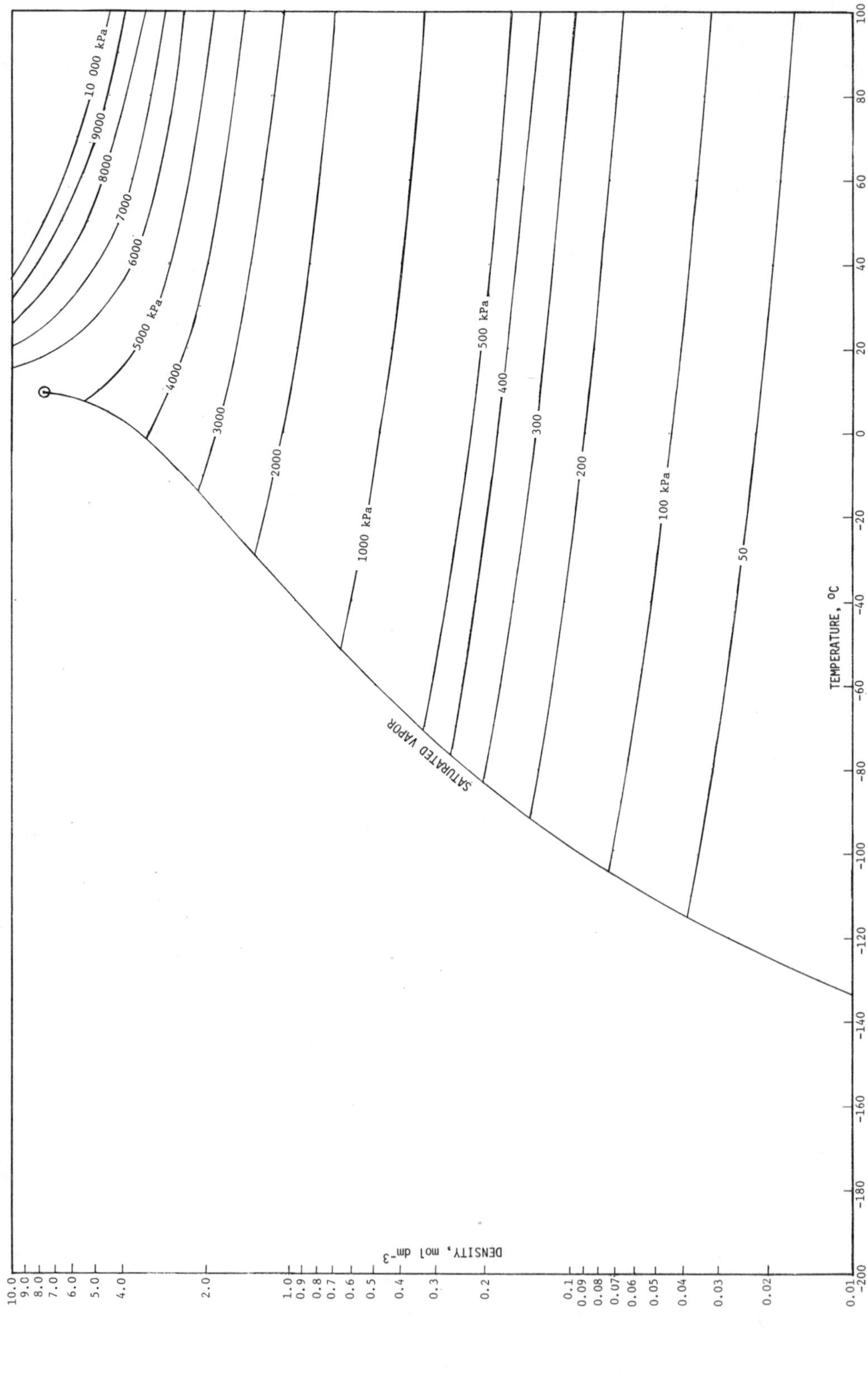

FIG. 2b. Density of ethylene.

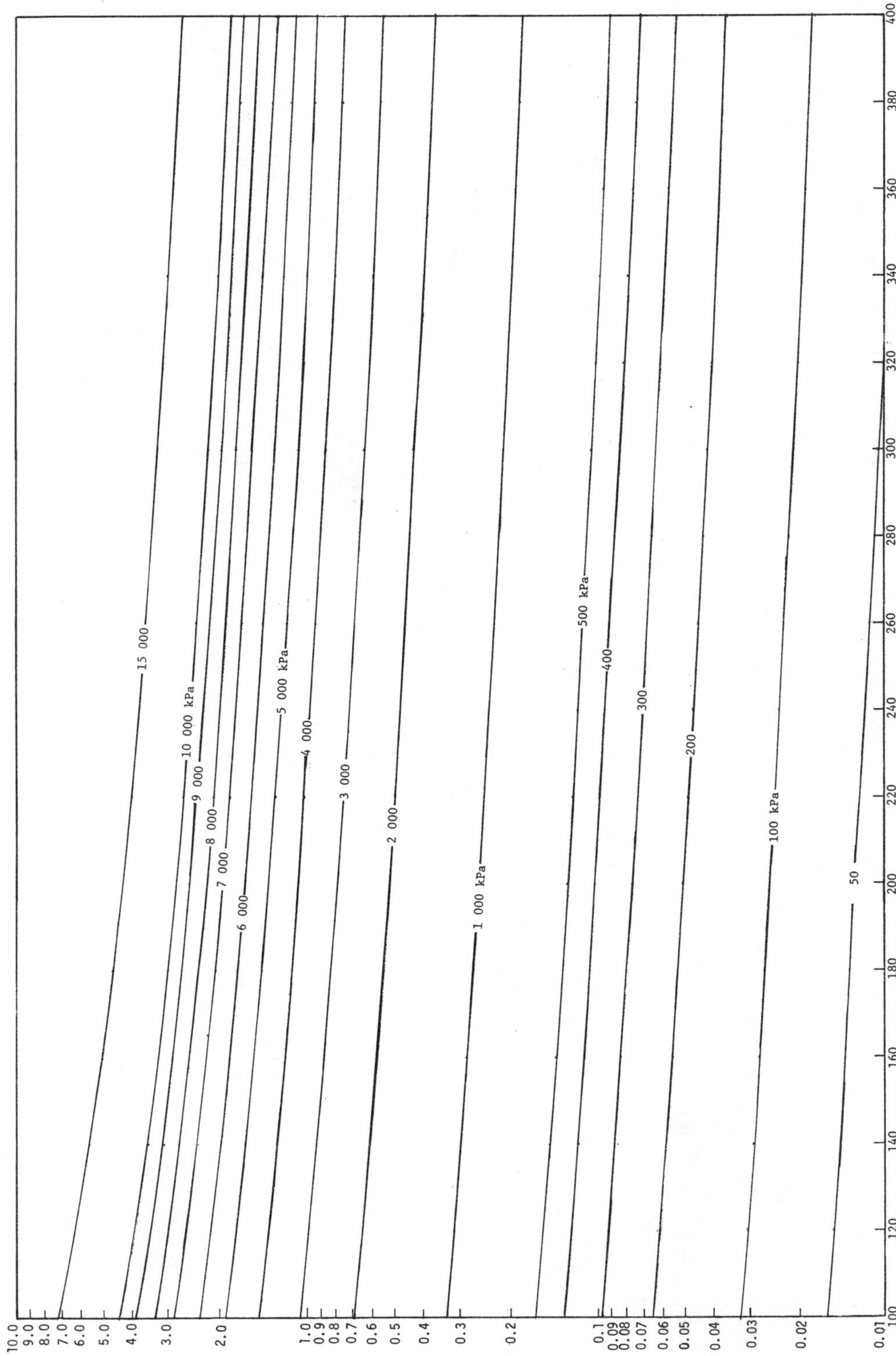

FIG. 2c. Density of ethylene.

FIG. 3a. Enthalpy of ethylene. Datum: Perfect crystal at −273.15°C.

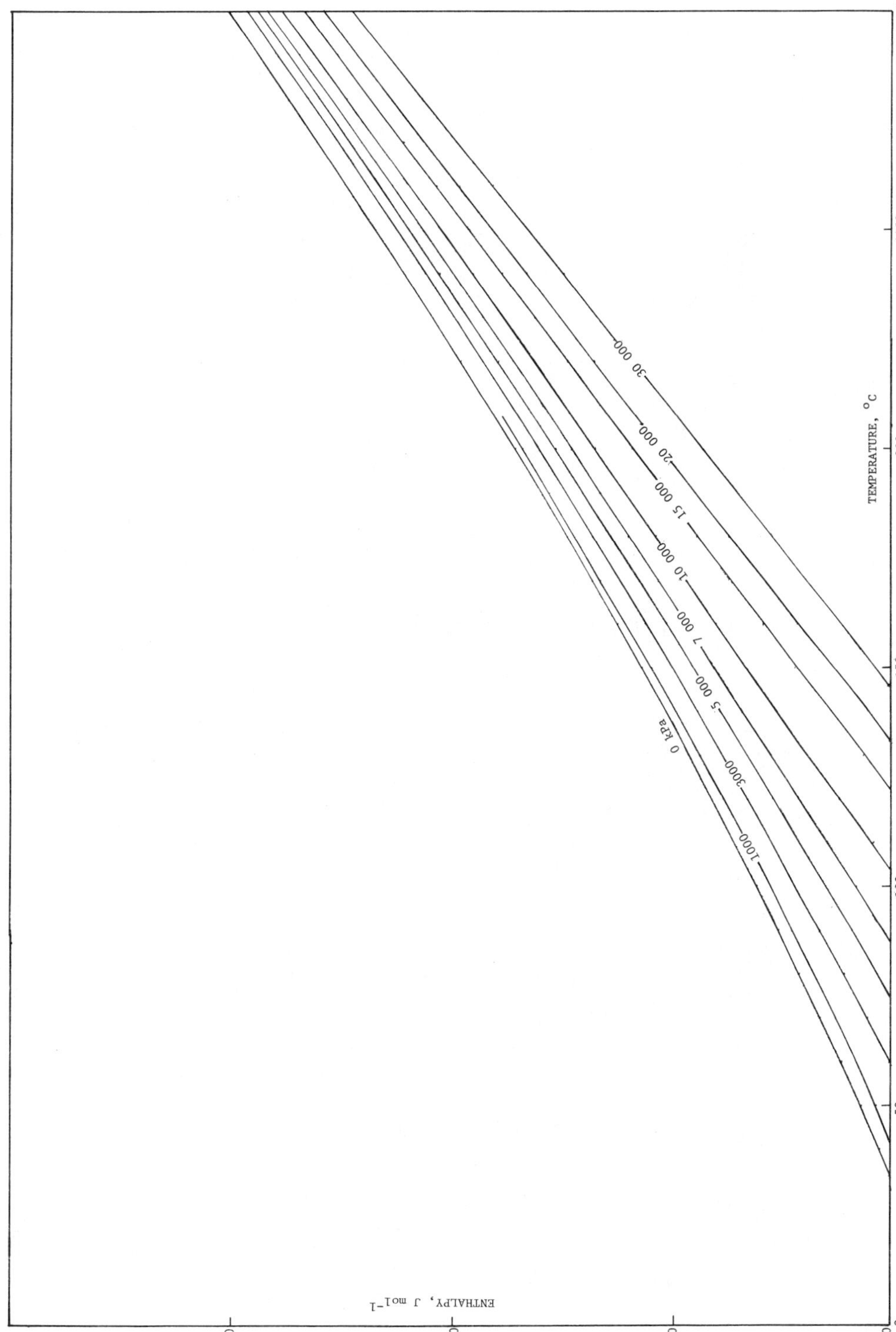

FIG. 3b. Enthalpy of ethylene. Datum: Perfect crystal at −273.15°C.

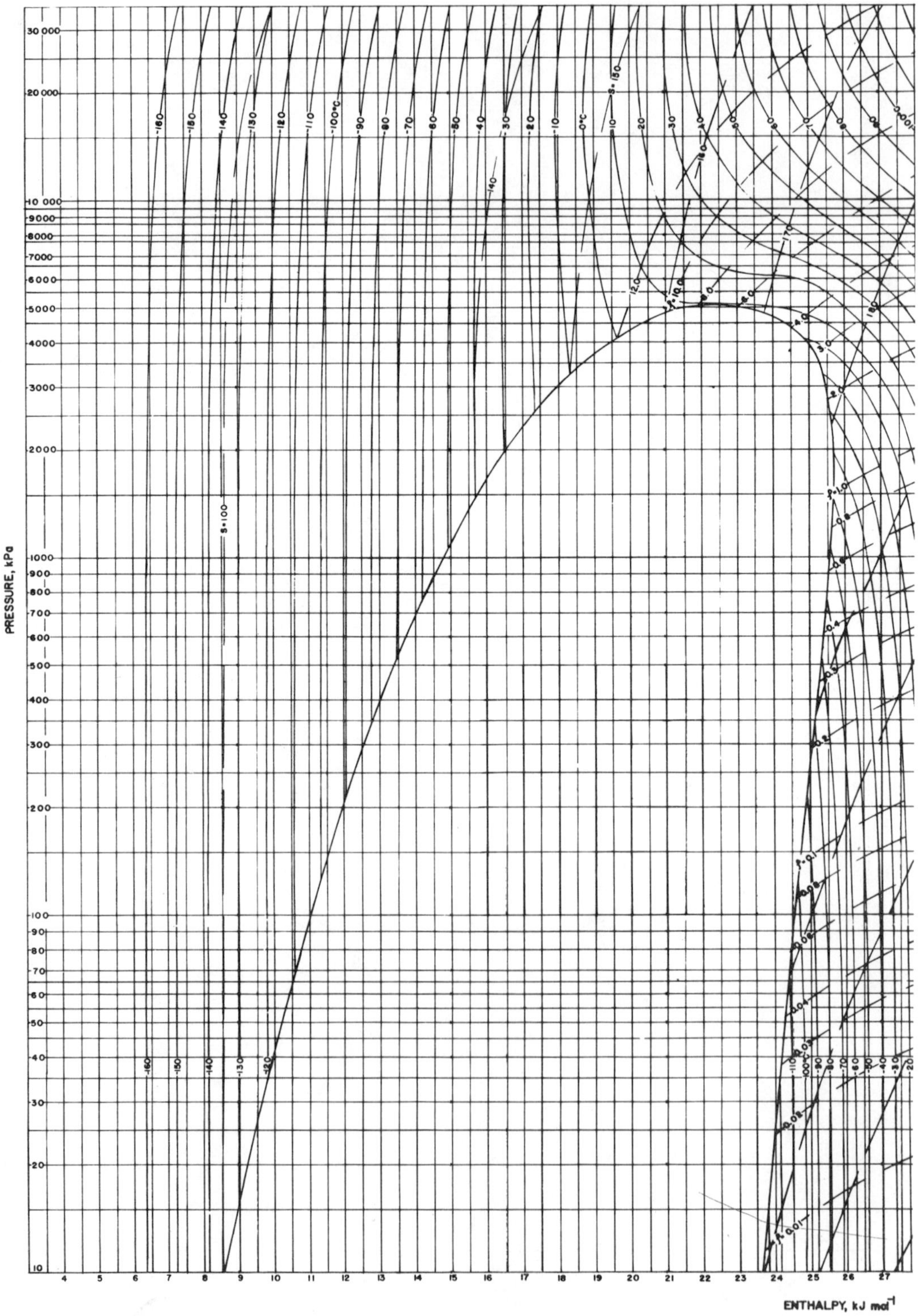
PRESSURE, kPa
30 000
20 000
10 000
9000
8000
7000
6000
5000
4000
3000
2000
1000
900
800
700
600
500
400
300
200
100
90
80
70
60
50
40
30
20
10
4
5
6
7
8
9
10
11
12
13
14
15
16
17
18
19
20
21
22
23
24
25
26
27
ENTHALPY, kJ mol⁻¹

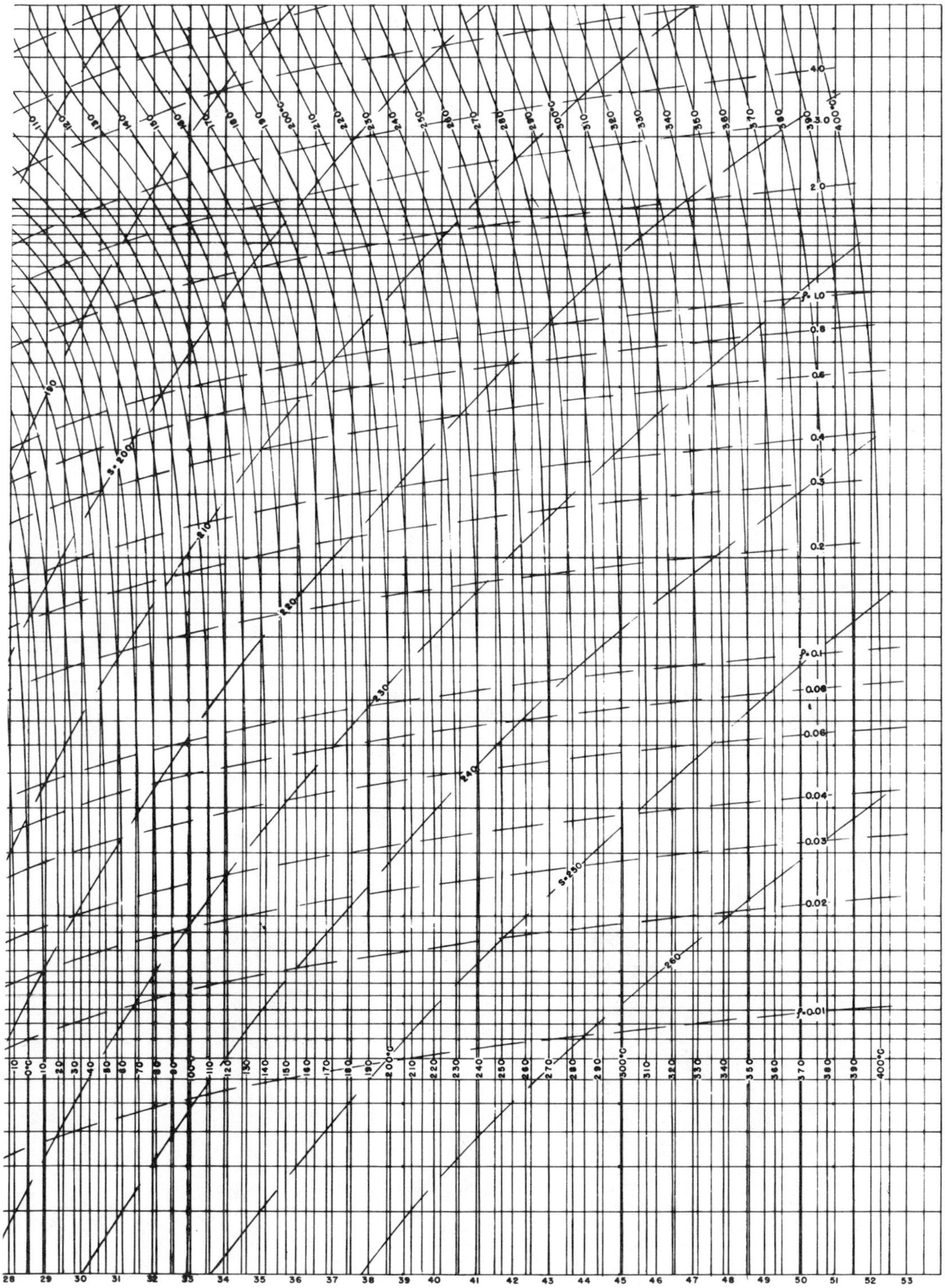

FIG. 4. Ethylene pressure–enthalpy diagram. T = temperature, °C. S = entropy, J mol^{-1} °K^{-1}. Density, mol/dm^3. Datum: perfect crystal at −273.15°C.

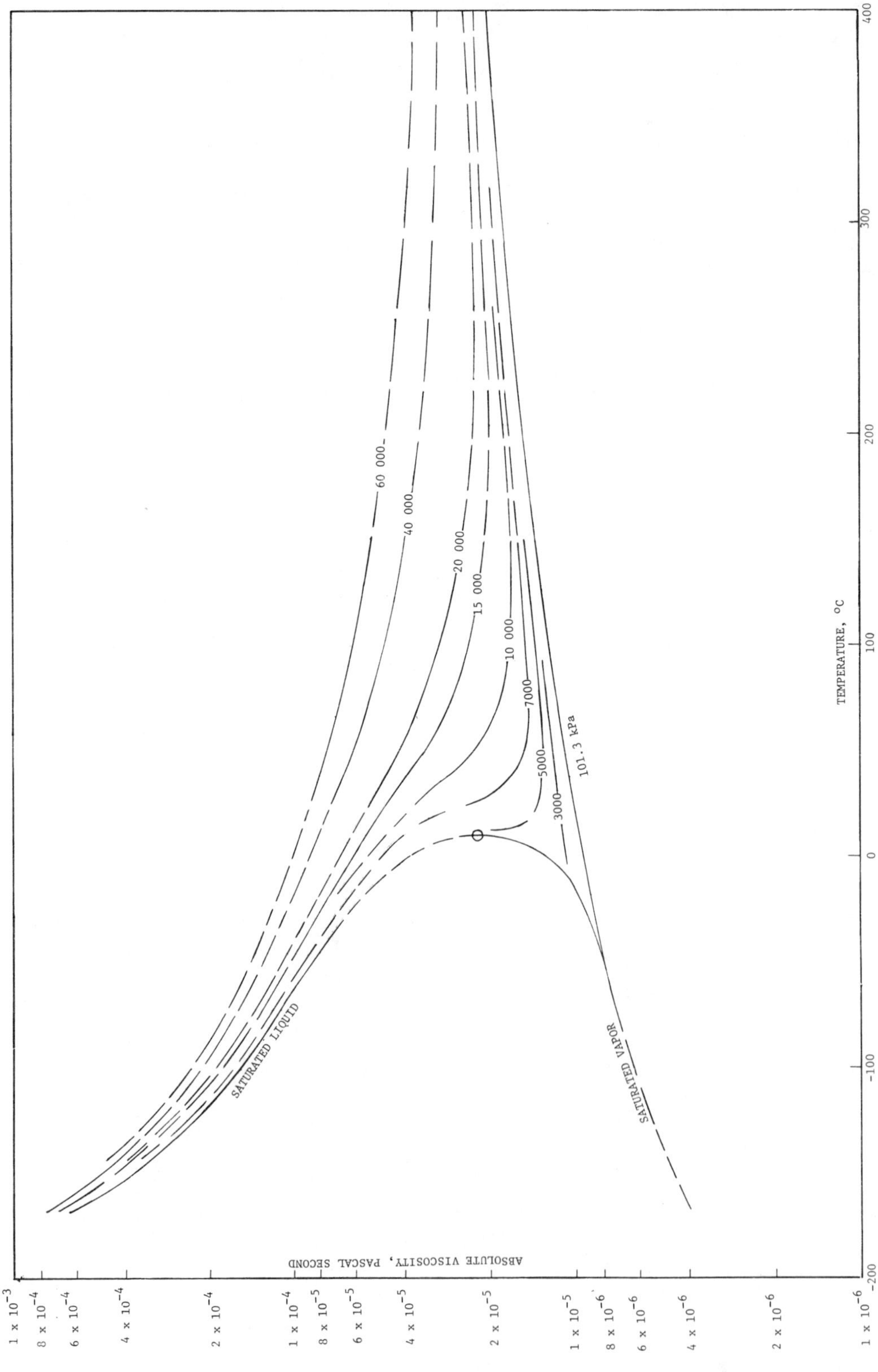

FIG. 5. Viscosity of ethylene.

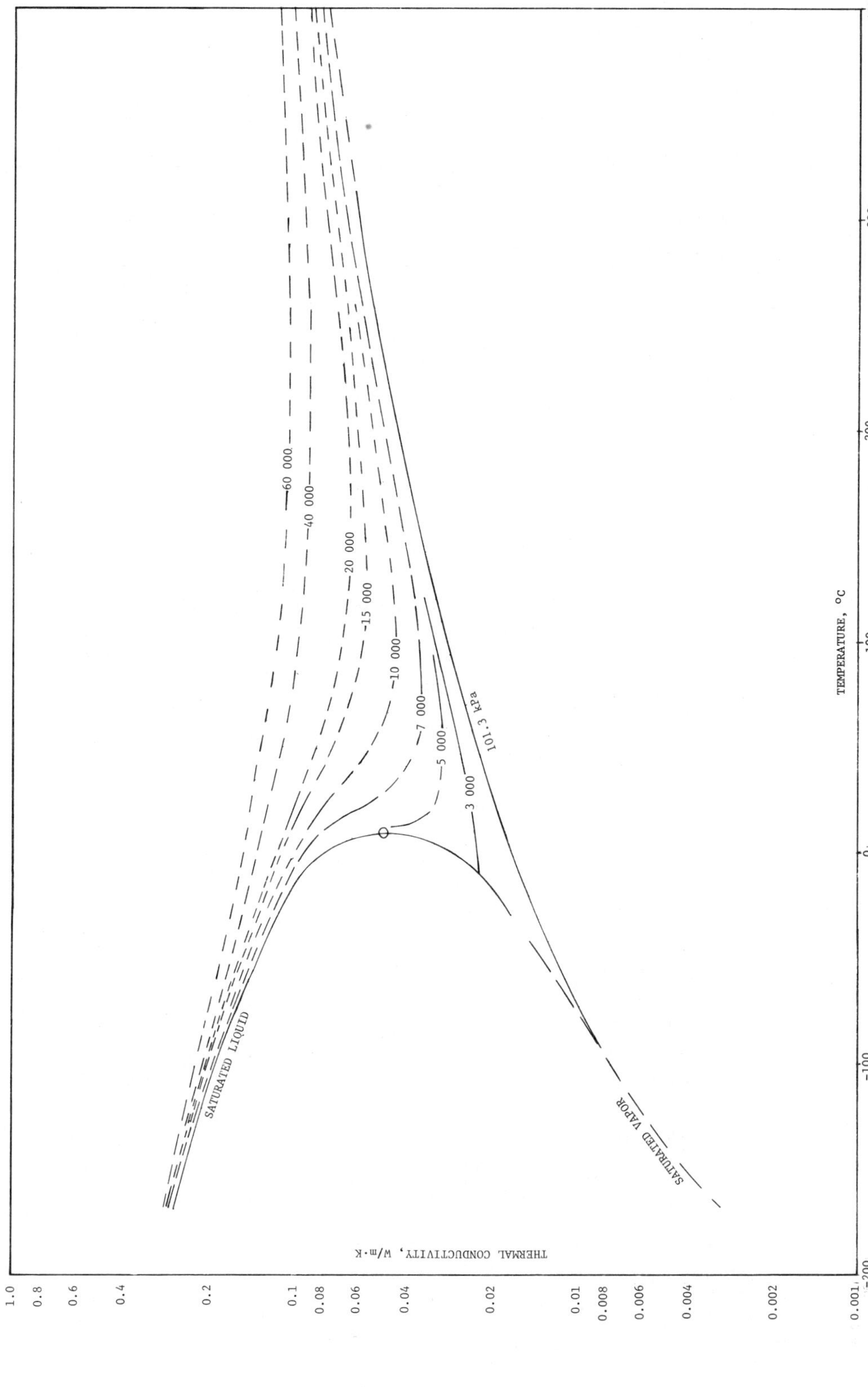

FIG. 6. Thermal conductivity of ethylene.

3. Chemical Reactions

Ethylene, due to its double bond, is a highly reactive compound which is converted to a multitude of intermediate and end products on a large industrial scale.

The variety of addition, oxidation, and polymerization reactions of industrial significance shows its reactivity. The discussion of chemical reactions of ethylene will be limited to the chemistry of processes for the production of first generation ethylene derivatives of major industrial importance. More details and additional references on individual processes utilizing chemical reactions of ethylene, on properties of first generation ethylene derivatives, and on descriptions of processes for the production of petrochemicals downstream of first generation ethylene derivatives may be found in the appropriate volumes of the *Encyclopedia of Chemical Processing and Design* [202]. The commercially important chemical reactions of ethylene are divided into seven basic reaction groups according to their similarity in chemical behavior. The reaction groups are arranged in the order of their industrial importance in the United States. Table 2 presents United States ethylene consumption in 1976 on this basis.

TABLE 2 United States Ethylene Consumption in 1976

Process Chemistry	Percent of Total U.S. Ethylene Consumption	First Generation Ethylene Derivatives	Percent of Total U.S. Ethylene Consumption
Polymerization	43	Polyethylene	42
		Ethylene–propylene elastomers	1
Oxidation	23	Ethylene oxide/glycol	18
		Acetaldehyde	3
		Vinyl acetate	2
Halogenation/ hydrohalogenation	16	Ethylene dichloride	15
		Ethyl chloride Ethylene dibromide Ethyl bromide	1
Alkylation	10	Ethylbenzene	9
		Ethyl toluene Aluminum alkyls Ethylanilines 1,4-Hexadiene	1
Hydration	4	Ethanol	4
Oligomerization	4	α-Olefins Linear primary alcohols	4
Oxo reactions	<1	Propionaldehyde	<1
	≅100		≅100

3.1. Polymerization

Polymers of ethylene represent, in terms of rate of production, the most important group of products of the petrochemical industry. The production of polyethylene ranks first in the order of ethylene consumption.

3.1.1. Polyethylene

In industrial production, polyethylene is formed by polymerization of so-called "polymerization grade" ethylene (99.9+% purity and virtually free of other olefins, acetylenes, and dienes) under specific conditions of temperature and pressure and in the presence of an initiator or catalyst:

$$n(\mathrm{CH_2{=}CH_2}) \longrightarrow (\mathrm{-CH_2-CH_2-})_n$$

The reaction is exothermic and may involve homogenous initiation (radical or cationic) or heterogenous initiation (solid catalyst) of the polymerization process. Polymers of a molecular weight of less than 1000 to 5 million or more (ultrahigh-molecular-weight polyethylene) may be formed. Although polyethylene is virtually defined by its name as a simple homopolymer of ethylene, the actual polymerization reactions and structures of the polyethylene molecules are complicated and diverse. Therefore, ethylene polymers are produced and are available as hundreds of compounds with widely different properties and structures, including copolymers, polymer mixtures, filled compositions, foamed products, and cross-linked products.

There are four types of basic reaction systems for the production of polyethylene that are of industrial importance:

High-pressure free-radical polymerization using oxygen, peroxides, or other strong oxidizers as initiator.

Low-pressure polymerization by heterogenous catalysis with catalysts of transition metal oxides such as molybdenum oxide or chromium oxide supported on inorganic carriers.

Low-pressure polymerization by anionic catalysis using Ziegler catalysts such as aluminum alkyls and titanium halides.

Low-pressure polymerization with Ziegler catalysts supported on inorganic carriers.

The high-pressure polymerization processes produce so-called low-density polyethylene (LDPE)—the older of the two polyethylenes—of a density between 0.910 and 0.940 g/cm^3. Oxygen, peroxides (e.g., *t*-butyl peroxide), or other strong oxidizers are used as polymerization initiators, and pressures between 60 and 350 MPa and temperatures of up to 350°C are applied. The processes utilized in industry today are basically similar to the process discovered by ICI in England in 1933 and use autoclaves or tubular reactors. The polymerization reaction is a typical free-radical initiation, polymer

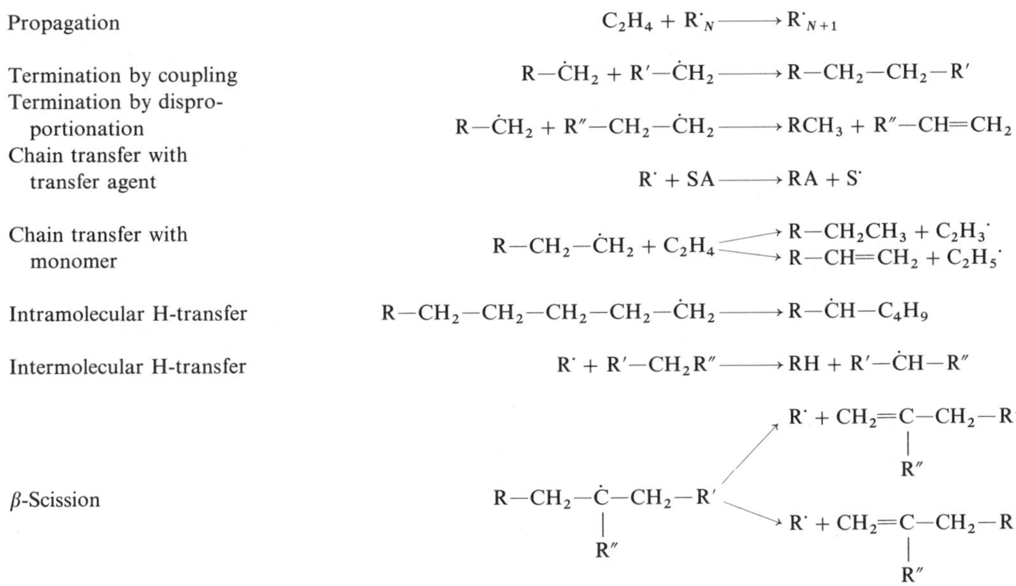

FIG. 7. Elementary reaction steps of radical ethylene polymerization.

propagation, and termination by combination. Several important side reactions involving chain transfer occur. Figure 7 [11] summarizes the most important steps of ethylene radical polymerization. Chain-transfer agents such as aldehydes, alkanes, hydrogen, ketones, and olefins are used to control molecular weight. The polymers formed are highly branched.

The conventional low-pressure processes produce so-called high-density polyethylene (HDPE) with a density of 0.941 to 0.970 g/cm^3. (The expression "medium-density polyethylene" is sometimes used for polyethylene with a density near 0.94 g/cm^3.) Densities close to 0.94 g/cm^3 are obtained by copolymerization with small amounts of other olefins. Catalysts vary widely, but usually fall into one of the three groups listed above. Polymerization temperatures range from 50 to 300°C at pressures of 0.1 to 20 MPa. (Sometimes processes of this type are divided into low- and medium-pressure processes.) The polymerization may be carried out in a fixed or fluid catalyst bed or with the catalyst slurried or suspended in the reactant mixture which may contain a solvent or diluent. Polymers produced by low-pressure processes are more linear in nature than those produced by high-pressure processes.

The most recent development in polyethylene technology is a vastly simplified low-pressure LDPE process developed as an extension of gas-phase fluid bed HDPE technology. The process employs a new catalyst system, pressures in the range of 0.7 to 2.1 MPa—as compared to 60 to 350 MPa in conventional LDPE technology—and temperatures of less than 100°C. Reaction is gas-phase instead of liquid-phase in conventional LDPE processes. Considerably lower capital and energy costs, less space requirements for production equipment, and enhanced safety are the advantages of the new

LDPE process [12]. Other low-pressure LDPE processes based on new catalyst technology are being explored by a number of LDPE producers [188].

New catalyst developments and improved end-product properties have led to a gradual tightening of ethylene specifications for polymerization purposes in recent years. HDPE is the most demanding end-use and sets the standards for a number of impurity concentration specifications. Generally, ethylene suitable for HDPE production is also satisfactory for LDPE production. A typical polymerization grade ethylene specification is given in Section 6.5. Further discussion of ethylene specifications is provided by Hahn, Chaptal, and Sialelli [13].

The amount of branching can be controlled in both LDPE and HDPE to adjust physical properties by controlling reaction conditions and by adding varying amounts of comonomers.

The most important structural features of polyethylene that influence its properties are:

Degree of branching in the polymer (determines crystallinity, density)
Average molecular weight (MW)
Molecular-weight distribution (MWD)

HDPE has fewer side-chain branches than LDPE and is highly crystallinic (90% or more). The crystallinity of LDPE is as low as 50%. LDPE is used primarily in the production of films, injection molding, wire, and cable and paper coatings; HDPE main uses are in production of blow- and injection-molded products.

Copolymers of ethylene with small amounts of another monomer have been formulated and are produced in significant commercial quantities to develop certain desired product qualities.

Olefins (propylene, 1-butene, 1-hexene, 1-octene) are copolymerized in small quantities in low-pressure processes to improve selected physical properties of HDPE as the density is reduced. Low-density copolymers containing varying amounts of ethyl acrylate (EEA), maleic anhydride (EMA), vinyl acetate (EVA), acrylic acid, and ionomers (having pendant carboxylate groups associated with mono- or divalent cations) are produced to modify selected physical properties over LDPE homopolymers. Strictly speaking, copolymers of this type are no longer polyethylene. However, if the amount of comonomer is small (as is the case in most of these polymers), they contain a backbone predominantly of polyethylene and are certainly closely related to it. Other commercially produced ethylene copolymers containing only small amounts of ethylene (e.g., vinyl acetate–ethylene copolymers with less than 25% ethylene or propylene copolymers with less than 5% ethylene) have different properties and should be discussed as a separate group. Their markets, however, are not yet fully developed and their ethylene consumption is not significant. Ethylene–propylene elastomers are discussed in the next section.

A detailed discussion of the polymerization mechanism, molecular structures, and properties of polyethylenes is given by R. A. V. Raff and of copolymers of ethylene by N. L. Zutty, T. A. Faucher, and S. Bonotto [14].

3.1.2. Ethylene–Propylene Elastomers

Ethylene–propylene elastomers possess outstanding resistance to oxygen, ozone, and heat, as well as excellent low-temperature flexibility, and are used mainly in automotive parts, tires, hoses, tubing, wires, and cables. The demand for these elastomers is not yet mature, and the present United States ethylene consumption for their product amounts to less than 1% of the total. Two types manufactured are copolymers (EP) by polymerizing ethylene and propylene, and terpolymers (EPT or EPDM) by polymerizing ethylene, propylene, and a small amount of suitable termonomers such as dienes and trienes (e.g., 1,4-hexadiene, dicyclopentadiene, 1,4,9-decatriene). The purpose of the termonomer is to make ethylene–propylene copolymers vulcanizable for traditional sulfur recipes by introducing double bonds in every macromolecule during polymerization. Terpolymers are produced in much larger volume than the copolymers. The polymerization of both is carried out either in the presence of an inert solvent (aliphatic, aromatic, or halogenated hydrocarbon) or in its absence (using liquid propylene as diluent) with a Ziegler-type catalyst (aluminum alkyl in the presence of hydrocarbon soluble vanadium compounds). A detailed discussion of ethylene–propylene co- and terpolymers is presented by A. Valvassori and N. Cameli [15].

3.2. Oxidation

Oxidation reactions of ethylene are of great industrial importance. Production of ethylene oxide/glycol ranks second, of acetaldehyde sixth, and of vinyl acetate eighth in the order of consumption of ethylene in the United States at the present time.

3.2.1. Ethylene Oxide/Glycol

Ethylene glycol ($HOCH_2CH_2OH$) is a derivative of ethylene oxide ($\overbrace{CH_2-CH_2}^{O}$) about two-thirds of which is consumed for the production of ethylene glycol. It is used mainly as antifreeze and in the production of polyester filament, fiber, and film. Ethylene oxide is also used for the production of surfactants, ethanolamines, glycol ethers, and higher glycols.

Ethylene glycol was at one time exclusively produced by the chlorhydrin process. In this process, hydrochlorous acid is added to ethylene at 25 to 50°C under 0.2 to 0.3 MPa pressure. The chlorohydrin is then treated with a caustic solution at 75°C:

$$2CH_2{=}CH_2 + 2HOCl \longrightarrow 2HOCH_2CH_2Cl \quad (1)$$

$$2HOCH_2CH_2Cl + Na_2CO_3 + H_2O \longrightarrow 2HOCH_2CH_2OH + CO_2 + 2NaCl \quad (2)$$

In another version of this process the chlorohydrin of Step (1) is recovered and then heated to 100°C with aqueous calcium hydroxide. Ethylene oxide is formed and separated from reaction salts by distillation and is then either hydrolized to form ethylene glycols or is used for other purposes:

$$2HOCH_2CH_2Cl + Ca(OH)_2 \longrightarrow 2\overset{O}{\overbrace{CH_2-CH_2}} + CaCl_2 + 2H_2O$$

The noncatalytic, high-pressure, high-temperature hydrolysis of ethylene oxide (typically at 230°C and 3.5 MPa) produces besides monoethylene glycol sizable amounts of higher glycols such as di- and triethylene glycol as by-products:

$$m\overset{O}{\overbrace{CH_2-CH_2}} + H_2O \longrightarrow HO(CH_2CH_2O)_mH$$

While the total conversion of ethylene oxide is virtually quantitative, selectivities of mono-, di-, and triethylene glycols are in the order of 88 to 96, 4 to 10, and 0.5 to 2%, respectively. The yield depends primarily on the dilution of the reactants.

The chlorohydrin route to ethylene oxide/glycol lost its importance as a commercial process for new facilities when the direct oxidation of ethylene became economically feasible. Today ethylene oxide is almost exclusively produced by vapor-phase oxidation of ethylene over selective silver oxide catalysts, typically at 200 to 300°C and 1.5 to 3.0 MPa pressure:

$$CH_2{=}CH_2 + \frac{1}{2}O_2 \longrightarrow \overset{O}{\overbrace{CH_2-CH_2}}$$

The selective catalysts minimize the conversion of ethylene to carbon dioxide and water.

Improvements in ethylene oxide synthesis are unlikely to go beyond 75 to 80% selectivity. The search for improved manufacture of ethylene glycol led, therefore, to the acetoxylation process which avoids the production of ethylene oxide as an intermediate. Ethylene is first converted to ethylene glycol mono- and diacetates catalytically and then hydrolized:

$$\begin{aligned} &4CH_2{=}CH_2 + 6CH_3COOH + 2O_2 \\ &\xrightarrow{\text{Catalyst}} 2CH_3COOCH_2CH_2OH + 2CH_3COOCH_2CH_2OOCCH_3 + 2H_2O \end{aligned} \qquad (3)$$

$$\begin{aligned} &2CH_3COOCH_2CH_2OH + 2CH_3COOCH_2CH_2OOCCH_3 + 6H_2O \\ &\longrightarrow 4HOCH_2CH_2OH + 6CH_3COOH \end{aligned} \qquad (4)$$

The acetic acid is recovered and recycled. A number of catalysts have been patented for this type of reaction with varying selectivities to ethylene glycol mono- and diacetates. A tellurium catalyst in conjunction with a bromine compound achieves the highest reported overall yield (about 93%), and a process using this catalyst system is being commercialized at the present time [189].

3.2.2. *Acetaldehyde*

The direct oxidation of ethylene is the preferred route to acetaldehyde (CH_3CHO) today. Prior to 1960, acetaldehyde was produced from ethanol, acetylene, and in smaller quantities from saturated hydrocarbons. It is used mainly for the production of acetic acid and acetic anhydride.

In the commercial direct oxidation process, three chemical reactions are utilized and combined into a workable liquid-phase processing scheme.

In this process, palladium chloride is reduced to elemental palladium which is reoxidized by cupric chloride. Oxygen converts the reduced cuprous chloride back to cupric chloride:

$$C_2H_4 + PdCl_2 + H_2O \longrightarrow CH_3CHO + Pd^0 + 2HCl \quad (5)$$

$$Pd^0 + 2CuCl_2 \longrightarrow PdCl_2 + 2CuCl \quad (6)$$

$$2CuCl + 2HCl + \frac{1}{2}O_2 \longrightarrow 2CuCl_2 + H_2O \quad (7)$$

$$C_2H_4 + \frac{1}{2}O_2 \longrightarrow CH_3CHO$$

The reaction proceeds at room temperature with yields of 93 to 95%.

The process can be carried out either in one or two stages. The three chemical reactions above are overall reactions and not individual kinetic steps. The detailed chemistry of the system is quite complicated. More information on this process and references on a discussion of the mechanisms involved may be found in the *Encyclopedia of Chemical Processing and Design*, Vol. 1 [190].

3.2.3. *Vinyl Acetate*

Prior to 1965, vinyl acetate ($CH_2CHOCOCH_3$) was produced either from acetylene or by cracking the reaction product of acetaldehyde and acetic anhydride. None of these feedstocks was derived from ethylene at that time.

The vapor-phase ethylene-based process is the preferred route to vinyl acetate today. It is used mainly for the production of polyvinyl acetate latices and resins, polyvinyl alcohol, and PVC copolymers.

The commercial production of vinyl acetate from ethylene, acetic acid, and oxygen started in 1965. Originally, a liquid-phase process was employed which caused considerable corrosion problems. The vapor-phase ethylene route employs palladium on carbon, alumina, or silica-alumina at about 175 to 200°C and 0.4 to 1.0 MPa as a catalyst:

$$CH_2{=}CH_2 + CH_3COOH + \frac{1}{2}O_2 \longrightarrow CH_2{=}CHOCOCH_3 + H_2O$$

The yield is 91 to 95% vinyl acetate based on ethylene.

3.3. Halogenation/Hydrohalogenation

Halogenation and hydrohalogenation reactions of ethylene are of considerable industrial significance. The most important petrochemical intermediate produced in this group is ethylene dichloride (EDC), which ranks third in United States ethylene consumption today and is the precursor of vinyl chloride monomer (VCM). Other intermediates of commercial interest in this group are ethyl chloride, ethylene dibromide, and ethyl bromide.

3.3.1. Ethylene Dichloride/Vinyl Chloride

Over 80% of the ethylene dichloride ($ClCH_2CH_2Cl$) produced in the United States is used for vinyl chloride (CH_2CHCl) manufacture which in turn is almost exclusively used for the production of polyvinyl chloride homopolymer and copolymer resins. Ethylene dichloride is also used for the production of trichloroethylene, perchloroethylene, methyl chloroform, vinylidene chloride, and ethyleneamines.

Originally, vinyl chloride was produced directly from acetylene and hydrogen chloride (HCl). The preeminent position of acetylene was successfully challenged by ethylene with the introduction of the balanced chlorination/oxychlorination process, the predominant route to vinyl chloride today. New VCM technology, however, based on ethane feedstock [16, 17] may affect ethylene consumption in the same manner as ethylene caused the demise of acetylene during the 1960s.

Ethylene dichloride may be made by the direct addition of chlorine to ethylene according to the reaction:

$$CH_2{=}CH_2 + Cl_2 \longrightarrow ClCH_2CH_2Cl \qquad \text{(chlorination)}$$

or it may be produced by reaction of ethylene, hydrogen chloride, and oxygen (in air):

$$CH_2{=}CH_2 + 2HCl + \frac{1}{2}O_2 \longrightarrow ClCH_2CH_2Cl + H_2O \qquad \text{(oxychlorination)}$$

The chlorination may be carried out in liquid- or vapor-phase reaction in the presence of a catalyst. Many alternative processes are known and vary mainly in the catalyst used and the conditions of reaction. Ferric chloride is predominantly used in commercial processes today. Chlorides of aluminum, antimony, or copper and ethylene dibromide have also been used as catalysts. Temperatures of 40 to 140°C are employed at low pressures (0.1 to 0.5 MPa). The conversions are high, and yields of 96 to 98% are obtained.

The oxychlorination is carried out in the vapor phase with a copper chloride catalyst supported on alumina silica or pumice at 200 to 350°C and typically at

1 MPa pressure in a packed or fluidized bed. The conversion of HCl is better than 95%.

If VCM is the desired product, the EDC is then cracked by pyrolysis at 450 to 515°C and elevated pressure (0.6 to 2.5 MPa) to yield VCM and HCl. The HCl is recycled to the oxychlorination reaction. A 50 to 70% conversion per pass and a 95 to 96% yield is obtained.

$$CH_2{=}CH_2 + Cl_2 \longrightarrow ClCH_2CH_2Cl \qquad \text{(direct chlorination)} \quad (8)$$

$$CH_2{=}CH_2 + 2HCl + \frac{1}{2}O_2 \longrightarrow ClCH_2CH_2Cl + H_2O \qquad \text{(oxychlorination)} \quad (9)$$

$$2ClCH_2CH_2Cl \longrightarrow 2CH_2{=}CHCl + 2HCl \qquad \text{(cracking)} \quad (10)$$

$$2CH_2{=}CH_2 + Cl_2 + \frac{1}{2}O_2 \longrightarrow 2CH_2{=}CHCl + H_2O$$

3.3.2. *Ethyl Chloride*

The production of ethyl chloride (CH_3CH_2Cl) is closely tied to tetraethyl lead. It is also used in small quantities as refrigerant, solvent, and ethylation agent. About 90% of ethyl chloride is produced by ionic addition of hydrogen chloride to ethylene in the presence of an aluminum chloride catalyst in a liquid-phase reaction carried out at 25 to 40°C:

$$CH_2{=}CH_2 + HCl \longrightarrow CH_3CH_2Cl$$

A dilute ethylene stream is used as feed, and the yield is approximately 90% of theoretical.

The remaining 10% of the production is obtained by catalytic reaction of hydrogen chloride with ethanol.

3.3.3. *Ethylene Dibromide*

Ethylene dibromide ($CH_2CH_2Br_2$), used in gasoline, antiknock fluids, and in vinyl bromide for flame retardants of fibers and plastics, is produced by the liquid-phase bromination of ethylene at 35 to 85°C and atmospheric pressure with ethylene dibromide as solvent:

$$CH_2{=}CH_2 + Br_2 \longrightarrow BrCH_2CH_2Br$$

The reaction is exothermic. The yield is approximately 90% of theoretical.

3.3.4. *Ethyl Bromide*

Ethyl bromide (CH_3CH_2Br), used mainly as an ethylating agent in the synthesis of pharmaceuticals and fine chemicals and as a solvent and refrigerant, can be obtained by reaction of ethanol with hydrobromic acid or with an alkalic bromide and sulfuric acid.

It can also be prepared from ethylene and hydrogen bromide by using a catalyst of anhydrous aluminum bromide:

$$CH_2CH_2 + HBr \longrightarrow CH_3CH_2Br$$

Another process achieves direct bromination of ethylene with hydrogen bromide using gamma-rays as the reaction promoter [191]. In commercial synthesis the gamma-rays are obtained from a cobalt-60 isotope source.

3.4. Alkylation

Alkylation reactions of ethylene play a major role in a number of important industrial processes. Ethylbenzene ranks fourth in United States ethylene consumption today. Aluminum alkyls, although only minor ethylene consumers in their production, are important as initiator and starter compounds in the production of α-olefins and linear primary alcohols, as catalysts in polymerization reactions, and in the synthesis of some monomers. Ethyl toluene, ethyl anilines, and 1,4-hexadiene are also of commercial interest.

3.4.1. Ethylbenzene

Ethylbenzene ($C_6H_5CH_2CH_3$) is used almost exclusively for the manufacture of styrene. Approximately 10% of the ethylbenzene produced in the United States is a by-product of refinery operations. The remaining 90% is produced synthetically via alkylation of benzene with ethylene:

$$C_6H_6 + C_2H_4 \longrightarrow C_6H_5C_2H_5$$

Most of the synthetic ethylbenzene is manufactured by means of the Friedel-Crafts reaction with aluminum chloride catalyst and a small amount of hydrochloric acid or ethyl chloride as initiator. Three phases are involved: ethylene gas, aromatic liquid, and a liquid heavy catalyst complex phase. The reaction takes place in the catalyst complex, and equilibrium is established between catalyst complex and organic phase. Both phases are subsequently separated and the catalyst complex is recycled to the alkylation reactor. The exact mechanism of the alkylation reaction is not fully understood, but is likely to involve first the protonization of the double bond of ethylene and then the attraction of this electrophilic reagent by the electrons of the benzene ring.

The alkylation reaction is complicated by side reactions such as cracking, polymerization, and hydrogen transfer. Of major importance is the formation of polyalkylated products such as di-, tri-, and higher ethylbenzenes which are converted to monoethylbenzene with benzene under suitable conditions in a separate reaction step via transalkylation (dealkylation) in the presence of $AlCl_3$ as catalyst. The catalyst used to alkylate reverses the alkylation reaction at elevated temperatures. The alkylation is carried out at temperatures of 80 to 100°C and pressures of 0.1 to 0.2 MPa; the transalkylation at temperatures of 200°C and higher.

An improved version using aluminum chloride eliminates the separate catalyst complex phase and thus operates at conditions resulting in a yield of over 99% and more reliable operation [192, 193].

Other catalysts used in alkylation of commercial ethylbenzene production are boron trifluoride supported on modified alumina, phosphoric acid on Kieselguhr, and silica–alumina. A new process using a solid, acidic zeolite catalyst is reported to provide a yield of over 99% [194, 195]. The alkylation reaction mechanisms with these catalysts are somewhat different.

3.4.2. *Ethyl Toluene*

Ethyl toluene ($C_2H_5C_6H_4CH_3$) is used mainly in the manufacture of vinyl toluene which in turn is utilized in certain types of unsaturated polyesters. The essential reactions involved are similar to those of styrene manufacture, i.e., alkylation of toluene with ethylene to form ethyl toluene by a Friedel-Crafts reaction followed by catalytic dehydrogenation. The alkylation is carried out with aluminum chloride catalyst at a temperature of 60 to 90°C and a pressure of about 0.4 MPa. A mixture of *m*- and *p*-ethyl toluene is formed. The overall yield of ethyl toluene is about 97%. Dehydrogenation of the mixture with zinc oxide catalyst produces a mixture of *m*- and *p*-vinyl toluene used mainly in the formulation of unsaturated polyester resins.

3.4.3. *Aluminum Alkyls*

Alkylation of aluminum with ethylene results in the formation of organoaluminum compounds which find wide commercial applications as initiator and starter compounds in other reactions, as polymerization catalysts, and in the synthesis of some monomers (e.g., 1,4-hexadiene).

Of the four ethylene-derived aluminum alkyls that are produced commercially (triethyl aluminum, diethyl aluminum chloride, ethyl aluminum dichloride, and ethyl aluminum sesquichloride), the most important is triethyl aluminum which is used for polymer catalysis and for the production of linear α-olefins and linear primary alcohols. Both trialkyl aluminum and alkyl aluminum halides are effective cocatalysts in Ziegler-type catalysts in conjunction with various transition metal halides for polymerization reactions primarily in low-pressure systems and stereospecific elastomer polymerization (e.g., in the production of polypropylene, high-density polyethylene, polybutadiene, polyisoprene, ethylene–propylene elastomers). Alkyl aluminum chlorides are also used as precusors for the production of trimethyl aluminum.

Several methods have been developed for the production of ethylene-based aluminum alkyls. They may be formed by reacting an aluminum alloy such as aluminum-magnesium with an ethyl halide at temperatures from 30 to 170°C, e.g.:

$$Al_2Mg_3 + 6C_2H_5Cl \longrightarrow 2Al(C_2H_5)_3 + 3MgCl_2$$

or by addition of ethylene to aluminum hydride:

$$AlH_3 + 3C_2H_4 \longrightarrow Al(C_2H_5)_3$$

An important industrial method starts with some previously prepared triethyl aluminum which reacts with aluminum, hydrogen, and ethylene in two steps:

$$2Al(C_2H_5)_3 + \frac{3}{2}H_2 + Al \longrightarrow 3Al(C_2H_5)_2H \tag{11}$$

$$3Al(C_2H_5)_2H + 3C_2H_4 \longrightarrow 3Al(C_2H_5)_3 \tag{12}$$

$$Al + \frac{3}{2}H_2 + 3C_2H_4 \longrightarrow Al(C_2H_5)_3$$

This route is often referred to as the "three for two" process since 2 mol of triethyl alumina are used to produce a total of 3 mol.

Another industrial method involves thermal displacement of other alkyl aluminum compounds with ethylene:

$$Al(R_1H)(R_2H)(R_3H) + 3C_2H_4 \longrightarrow Al(C_2H_5)_3 + R_1 + R_2 + R_3$$

R_1, R_2, and R_3 are oligomers of ethylene of the same or varying chain length. The displacement reaction is utilized as a recovery step in the manufacture of α-olefins from long-chain aluminum alkyls and to convert other commercial aluminum alkyls (e.g., triisobutyl aluminum, diisobutyl aluminum hydride) to triethyl aluminum.

Industrial production of alkyl aluminum chlorides is carried out by first reacting aluminum with ethyl chloride to produce ethyl aluminum sesquichloride:

$$2Al + 3C_2H_5Cl \longrightarrow Al_2(C_2H_5)_3Cl_3$$

The individual chlorides are then made by adding the correct amount of triethyl aluminum or aluminum chloride to manufacture diethyl aluminum chloride or ethyl aluminum dichloride, respectively:

$$Al_2(C_2H_5)_3Cl_3 + Al(C_2H_5)_3 \longrightarrow 3Al(C_2H_5)_2Cl$$
$$Al_2(C_2H_5)_3Cl_3 + AlCl_3 \longrightarrow 3Al(C_2H_5)Cl_2$$

These compounds can also be produced directly from triethyl aluminum and aluminum chloride by using stoichiometric quantities for the reactions:

$$2Al(C_2H_5)_3 + AlCl_3 \longrightarrow 3Al(C_2H_5)_2Cl$$
$$2Al(C_2H_5)_3 + 4AlCl_3 \longrightarrow 6Al(C_2H_5)Cl_2$$

More information and references on aluminum alkyls may be found in the *Encyclopedia of Chemical Processing and Design*, Vol. 3 [203].

3.4.4. Ethylanilines

Ethylanilines are used as intermediates in the manufacture of dyes, pharmaceuticals, and pesticides.

Ethylanilines are formed by direct alkylation of aniline with ethylene under heat and pressure and with various catalysts. Aluminum aniline, for example, catalyzes the reaction in such a way that a mixture of 2-ethyl and 2,6-diethyl aniline is formed. Under certain conditions as much as 90% of the diethyl derivative can be obtained.

Another method of preparing ethylanilines is to react ethanol and aniline in the presence of sulfuric acid.

3.4.5. 1,4-Hexadiene

1,4-Hexadiene ($CH_2CHCH_2CHCHCH_3$) is used as a third monomer in ethylene–propylene terpolymer manufacture and is produced by reacting ethylene and butadiene in the presence of a Ziegler-type catalyst:

$$CH_2CHCHCH_2 + CH_2CH_2 \longrightarrow CH_2CHCH_2CHCHCH_3$$

3.5. Hydration

The hydration products of ethylene are ethanol and diethyl ether, both of which have had considerable industrial importance for many years and are still important as chemical intermediates and products.

Although ethyl ether can be made from ethanol via sulfuric acid dehydration, most of it is obtained as a by-product from synthetic ethanol units. In both cases ethanol is a precursor of diethyl ether.

3.5.1. Ethanol

The process chemistry of ethanol (C_2H_5OH) production has changed remarkably during the past four decades. In the 1930s almost all industrial ethanol was produced by fermentation in the United States. Today, less than 5% is made by this route and over 95% is synthesized from ethylene. Synthetic ethanol's share of United States ethylene consumption has fallen from about 22 to 4–6% in the last 15 years, not due to more economical routes for ethanol manufacture, but because of competing routes to some of ethanol's principal conversion products (acetaldehyde, acetic acid, acetate solvents). Other uses of ethanol are in toiletries, cosmetics, industrial solvents and thinners, detergents, flavors, disinfectants, food and drug processing, pharmaceutical products, and in the manufacture of amines and ethyl halides.

Industrial ethanol used to be produced from sugar-containing materials such as molasses by fermentation utilizing yeast enzymes to convert the sugar into ethanol. There is renewed interest in this route to ethanol in many developing countries that have scant petroleum and very limited capital resources (see Section 6.6.2).

Impending petroleum shortages may also make fermentation ethanol again attractive to industrialized countries. An advantage is that the raw materials are renewable. Any fermentable sugar can be used as feedstock. Recent developments in acid and enzymatic hydrolysis of cellulose to fermentable sugars may possibly allow the manufacture of fermentation ethanol as chemical feedstock and/or fuel economically from various cellulose materials which are renewable and available in vast quantities in nature and as waste products of civilization.

Ethylene can be converted to ethanol by way of ethyl sulfate or by direct hydration. The older of the two major synthetic routes to ethanol is the esterification-hydrolysis or absorption process. Ethylene is absorbed in 90 to 98% sulfuric acid at a reaction temperature of 50 to 85°C and a pressure of 1.0 to 1.4 MPa. The absorbate containing mixed sulfates is hydrolyzed with water at about 70°C to ethanol, dilute sulfuric acid, and small amounts of diethyl ether which is a by-product. The diluted sulfuric acid is separated from the product stream and reconcentrated. The reactions involved are as follows:

Absorption:

$$CH_2{=}CH_2 + H_2SO_4 \longrightarrow \underset{\text{Ethyl sulfate}}{CH_3CH_2OSO_3H}$$

$$2CH_2{=}CH_2 + H_2SO_4 \longrightarrow \underset{\text{Diethyl sulfate}}{(CH_3CH_2)_2SO_4}$$

Hydrolysis:

$$CH_3CH_2OSO_3H + H_2O \longrightarrow CH_3CH_2OH + H_2SO_4 \quad \text{(diluted)}$$

$$(CH_3CH_2)_2SO_4 + 2H_2O \longrightarrow 2CH_3CH_2OH + H_2SO_4 \quad \text{(diluted)}$$

$$(CH_3CH_2)_2SO_4 + CH_3CH_2OH \longrightarrow CH_3CH_2OSO_3H + \underset{\text{Diethyl ether}}{CH_3CH_2OCH_2CH_3}$$

Small amounts of diethyl ether ($C_2H_5OC_2H_5$) are formed as by-product:

$$2C_2H_5OH \longrightarrow C_2H_5OC_2H_5 + H_2O$$

The overall yield of ethanol based on ethylene is about 88% by weight. This process, however, is of declining industrial importance and almost all synthetic ethanol is now being produced by the direct hydration of ethylene over phosphoric acid or tungsten-based catalysts. Typically, in this process a highly concentrated ethylene stream (98+% ethylene) is compressed to about 7 MPa, mixed with water, vaporized, then further heated to a reaction temperature of 300°C and passed over a phosphoric acid catalyst absorbed on diatomaceous earth. The ethylene is directly hydrated to ethanol:

$$CH_2CH_2 + H_2O \longrightarrow CH_3CH_2OH$$

The reaction is exothermic. Traces of acetaldehyde and some ethyl ether are by-products of the reaction. The ethyl ether may be recycled. The conversion rate of ethylene is controlled by temperature, pressure, and the water to ethylene mole ratio. The product yield is about 95%.

3.6. Oligomerization

α-Olefins and linear primary alcohols in the C_4 to C_{20} range produced by the oligomerization of ethylene have become important industrial chemicals in recent years, especially for the production of plasticizers (lower range alcohols, C_6 to C_{10}) and biodegradable surface-active agents (higher range alcohols, C_{12} and up), but also in smaller quantities for the manufacture of synthetic lubricants, fatty amine oxides, thermoplastic comonomers, synthetic fatty oxides, and other products of minor importance. α-Olefins are precursors of plasticizer- and detergent-range alcohols.

3.6.1. α-Olefins

Linear α-olefins are produced commercially by the cracking of normal paraffin waxes and by the oligomerization of ethylene.

The synthesis of α-olefins from ethylene is based on Karl Ziegler's ethylene chain-growth chemistry work in the 1950s. It comprises the addition of ethylene to an aluminum alkyl compound, i.e., triethyl aluminum, which is first synthesized by reductive alkylation of aluminum and serves as a catalyst system wherein the ligands undergo oligomerization. When a certain ligand chain length is achieved, the alkyl group is displaced by an ethylene molecule. The growth and displacement reactions can be concurrent or successive.

$$Al + \frac{3}{2}H_2 + 2Al(C_2H_5)_3 \longrightarrow 3Al(C_2H_5)_2H \quad (13)$$

$$3Al(C_2H_5)_2H + 3C_2H_4 \longrightarrow 3Al(C_2H_5)_3 \quad \text{(triethylaluminum synthesis)} \quad (14)$$

$$Al(C_2H_5)_3 + nC_2H_4 \longrightarrow AlR_1R_2R_3 \quad \text{(growth reaction)} \quad (15)$$

$$AlR_2R_3{-}CH_2CH_2R + C_2H_4 \longrightarrow AlR_2R_3{-}C_2H_5 + CH_2{=}CHR \quad \text{(displacement reaction)} \quad (16)$$

R, R_1, R_2, and R_3 are oligomers of ethylene of the same or different chain lengths. All alkyl groups grow, and their average molecular weight can be controlled to a certain extent by the amount of ethylene added. At temperatures of 80 to 120°C and at pressures of 20 MPa the displacement reaction is very slow but the chain-growth reaction proceeds at an acceptable rate. This permits α-olefin synthesis to be conducted in two stages, an initial chain-growth reaction at 80 to 120°C and 20 MPa (high ethylene pressure favors growth reaction) followed by a later displacement reaction at 245 to 300°C and 0.7 to 2.0 MPa. The chain-growth and displacement reactions can also be carried out in one step at 160 to 275°C and 13.5 to 27.0 MPa.

The olefin-product distribution from the two-step process is described by a Poisson distribution; from the one-step process by a somewhat different distribution.

3.6.2. Linear Primary Alcohols

Higher molecular weight alcohols were first produced in commercial quantities from natural oils and fats such as coconut oil and tallow, the prices of which fluctuated widely. Since the early 1960s synthetic, more economical routes have been developed for the manufacture of alcohols similar or identical to the naturally derived products.

Higher aliphatic alcohols fall into three broad groups. The first group is linear primary alcohols which are based on ethylene and discussed in this section. The second group consists of primary alcohols with moderate branching which are produced by oxo reactions of linear olefins, and the third group is secondary alcohols produced by oxidation of normal paraffins.

Lower range linear primary alcohols (C_6 to C_{10}) are used in the manufacture of plasticizers, particularly for polyvinyl chloride. Higher linear primary alcohols (C_{12} and up) are mainly used in the form of derivatives (e.g., alcohol sulfates, ethoxylates, or sulfated ethoxylates) as surfactants for detergents. They compete in these areas with other higher aliphatic alcohols.

The synthesis of linear primary alcohols ($CH_3CH_2[CH_2CH_2]_nOH$) from ethylene begins in the same way as the synthesis of ethylene-derived α-olefins, i.e., with the synthesis of triethyl aluminum from aluminum, hydrogen, and ethylene. Two-thirds of the triethyl aluminum produced is recycled:

$$Al + \frac{3}{2}H_2 + 2Al(C_2H_5)_3 \longrightarrow 3Al(C_2H_5)_2H \quad (17)$$

$$3Al(C_2H_5)_2H + 3C_2H_4 \longrightarrow 3Al(C_2H_5)_3 \quad (18)$$

High purity ethylene is added in the next step to form higher alkyl aluminum compounds (growth reaction). Then the trialkyl aluminum is oxidized with air to form alkoxides which in turn are hydrolized to form straight-chain alcohols with an even number of carbon atoms:

$$Al(C_2H_5)_3 + nC_2H_4 \longrightarrow AlR_1R_2R_3 \quad (19)$$

$$AlR_1R_2R_3 + \frac{3}{2}O_2 \longrightarrow Al(R_1O)(R_2O)(R_3O) \quad (20)$$

$$Al(R_1O)(R_2O)(R_3O) + 3H_2O \longrightarrow R_1OH + R_2OH + R_3OH + Al(OH)_3 \quad (21)$$

R_1, R_2, and R_3 are oligomers of ethylene of the same or different chain length. Hydrolysis with excess sulfuric acid produces by-product aluminum sulfate. Water hydrolysis followed by drying and calcining yields a catalytic grade alumina.

3.7. Oxo Reaction (Hydroformylation)

The oxo reaction is the formation of an aldehyde from an olefin and synthesis gas with a hydrogen/carbon monoxide ratio of 1:1 or 2:1 over a cobalt catalyst at a temperature of 60 to 200°C and a pressure of 4 to 35 MPa:

$$RCH{=}CH_2 + CO + H_2 \longrightarrow RCH_2CH_2CHO$$

The cobalt catalyst is separated and the aldehyde reduced with hydrogen to produce an alcohol, typically at 150°C and 10 MPa in the presence of a nickel or copper chromate catalyst. Alcohols can also be synthesized from olefins in one step with phosphine-complexed cobalt catalysts:

$$RCH{=}CH_2 + CO + 2H_2 \longrightarrow RCH_2OH$$

Several processes using rhodium catalysts are in operation. Rhodium catalysts require lower pressure (0.1 to 5 MPa).

Although any olefin including ethylene can be used as feedstock to an oxo alcohol plant, the most important olefin raw material for oxo alcohols to be used in plasticizers and as surfactants is, for economic reasons, propylene.

3.7.1. Propionaldehyde

Propionaldehyde (CH_3CH_2CHO), however, is produced economically via the oxo reaction from ethylene. The derivatives propionic acid and *n*-propyl alcohol are both primarily used in agricultural applications, solvents, and chemicals.

$$C_2H_4 + CO + H_2 \longrightarrow CH_3CH_2CHO$$

$$CH_3CH_2CHO + H_2 \longrightarrow CH_3CH_2CH_2OH$$

$$CH_3CH_2CHO + \frac{1}{2}O_2 \longrightarrow CH_3CH_2COOH$$

3.8. Conclusion

Worldwide demand for ethylene-based petrochemicals is projected to continue its growth through the remainder of this century. Several potential new routes to ethylene-based intermediates and new end uses of ethylene of commercial significance could develop. These may involve chemical reactions of ethylene already known and/or new reaction systems with new compounds and catalysts and/or extreme reaction conditions.

4. Uses and Markets

All of the reactions described in the preceding chapter are practiced on an industrial scale in countries with highly developed chemical industries. Although a few unusual uses of ethylene have been developed, such as its application in the ripening of various fruits, most ethylene is consumed by the chemical industry which converts this raw material into a variety of products. The "tree of derivatives" is shown in Fig. 8 which illustrates the significant variety of organic intermediates and products derived from ethylene. The ethylene derivatives, products, and end uses shown are limited to those of commercial importance. The commercially important chemical reactions of ethylene are organized in seven basic reaction groups according to their chemistry. The reaction groups are arranged from the top of the figure to its bottom in the order of their industrial importance in the United States. The more important petrochemical intermediates, petrochemical products, and end uses are then shown in their relation to these groups of basic chemical reactions of ethylene.

In general, petrochemical intermediates are defined here as reaction products of ethylene which are processed further chemically in downstream petrochemical or chemical facilities. Petrochemical products, on the other hand, are not processed any further chemically after their production and are sold to producers of end use products (consumer and industrial products) who may use these directly (e.g., as solvents, additives) or who may process them further physically or mechanically (e.g., blending, molding):

$$\text{Ethylene} \longrightarrow \text{petrochemical intermediates} \longrightarrow \text{petrochemical products} \longrightarrow \text{end uses}$$

e.g.,

$$\text{Alkylation of ethylene} \longrightarrow \text{ethylbenzene} \longrightarrow \text{styrene} \longrightarrow \text{polystyrene} \longrightarrow \text{packaging materials}$$

These definitions and classification of ethylene derivatives into intermediates, products, and end uses are necessarily arbitrary. Many intermediates may also represent products, e.g., ethanol may be used as the precursor for other intermediates such as acetaldehyde and ethylamines, or for products such as ethyl acetate, or it may directly find end use as a solvent and disinfectant.

No world market for ethylene presently exists because ethylene is not easily transported across oceans. Consequently, three large separate and several small markets have developed over the last two decades in three distinct geographic areas: the American Gulf Coast, Western Europe, and Japan. The first two are characterized by extensive pipeline systems connecting a large number of producers and consumers, while in Japan the trading of ethylene between various petrochemical centers is carried out in the liquid state with relatively small specialized cryogenic tankers.

Smaller markets exist in Italy, Spain, the American Midwest, and the Comecon countries of Eastern Europe. Plans have been under discussion to connect American midwestern producers and consumers with Canadian plants in Alberta and Ontario. Similar plants exist for producers and consumers in several Comecon countries. In 1978 it remains an open question to what extent the abundant availability of feedstocks in the oil-producing countries will spur any worldwide trading in ethylene.

The distribution of existing and projected worldwide ethylene production [18–20] is shown in Table 3.

The projected worldwide growth rate, although lower than in the last 10 years, exceeds the projected growth rate of the total of the world's GNP which is estimated to amount to approximately 4%/yr in real terms. The low growth rate projection for capacity addition in Japan reflects the current underutilization of its existing capacity as well as its growing interest in joint-venture participation in petrochemical projects outside of Japan, notably in developing countries of the Far East, Southeast Asia, and the Middle East.

The table at least qualitatively reflects the vigorous growth of petrochemical industries in Latin America, the Far East, the Middle East, and Eastern Europe. Some of this growth will be at the expense of growth in the industrialized countries.

No reliable statistics exist on that portion of ethylene capacity which is dedicated to captive consumption by the producers. When ethylene is traded it is generally under long-term contracts; spot trading is rare.

TABLE 3 Worldwide Ethylene Production Capacity[a]

	1975 (10^6 t/yr)	1980 (10^6 t/yr)	Growth Rate (%/yr)
Japan	5.0	5.7	2.7
United States	10.8	17.9	10.6
Western Europe	12.7	17.8	7.0
Other areas	5.6	11.5	15.4

[a]In metric tons (t).

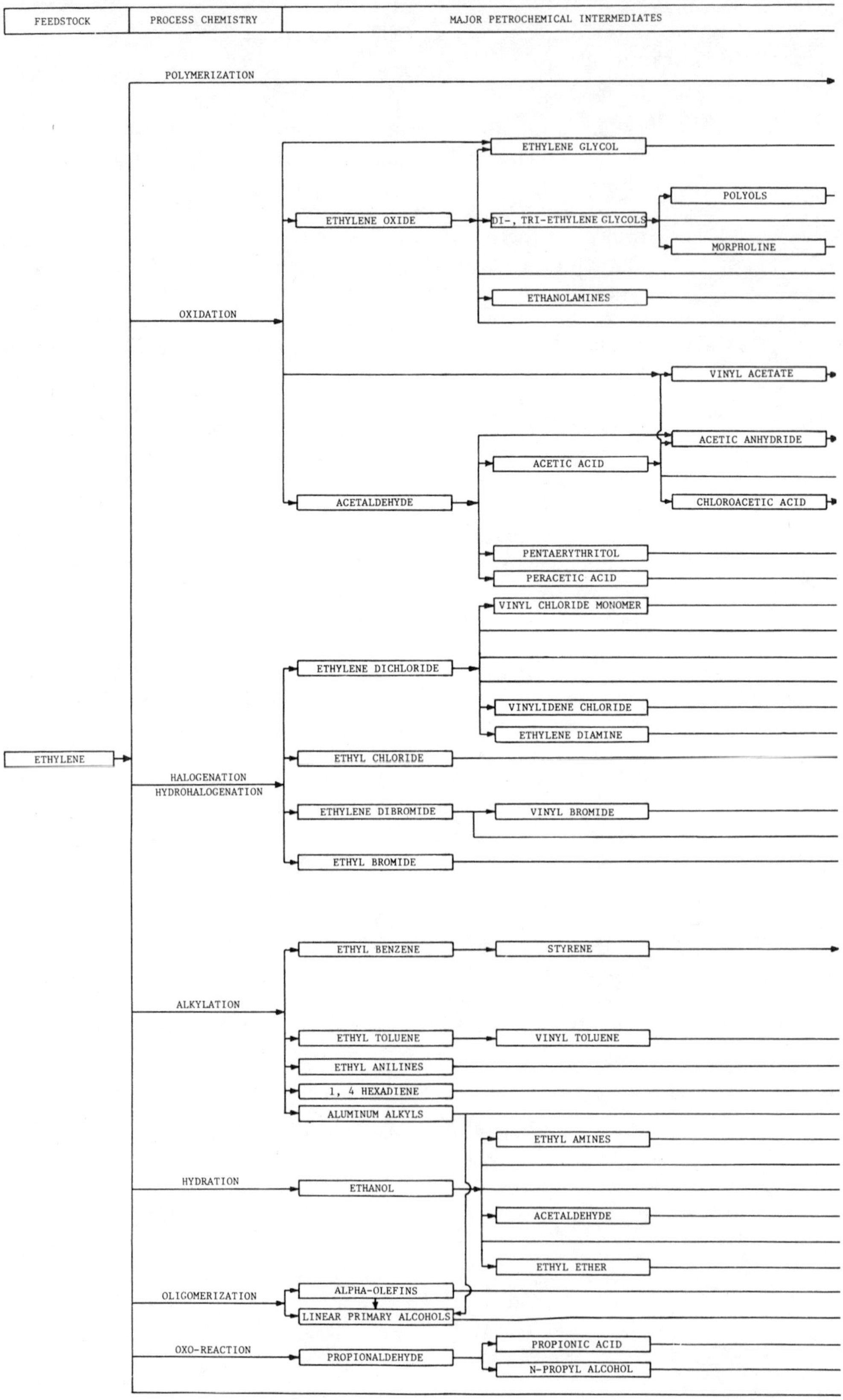

FIG. 8. Major ethylene

MAJOR PETROCHEMICAL PRODUCTS	MAJOR END USES
POLYETHYLENE, LOW-DENSITY	FILM & SHEET; INJECTION MOLDING; PAPER, WIRE & CABLE COATINGS
POLYETHYLENE, HIGH-DENSITY	BLOW MOLDING; INJECTION MOLDING; PIPE, CONDUIT, TUBING; FILM & SHEET
POLYETHYLENE COPOLYMERS	FILM; ELECTRICAL INSULATIONS; HOT MELT COATINGS & ADHESIVES
ETHYLENE-PROPYLENE CO- & TERPOLYMERS	AUTOMOTIVE PARTS; TIRES; HOSE & TUBING, THERMOPLASTIC MODIFIERS; WIRE & CABLE
	ANTIFREEZE
POLYESTERS	FILAMENT; FIBER; FILM
POLYURETHANE & POLYESTERS	FILAMENT; FIBER; FILM; FLEXIBLE & RIGID FOAMS; COATINGS; ELASTOMERS; ADHESIVES; SEALANTS
	DEHYDRATION AGENT; SOLVENTS
	RUBBER CHEMICALS; CORROSION INHIBITORS; OPTICAL BRIGHTENERS
GLYCOL ETHERS	SOLVENTS
	LIQUID DETERGENTS; ACID GAS ABSORBENT; TEXTILE CHEMICALS
	SURFACTANTS
POLYVINYL ACETATE	EMULSION PAINTS; ADHESIVES; TEXTILE CHEMICALS; PAPER & SURFACE COATINGS
POLYVINYL ALCOHOL	ADHESIVES; TEXTILE CHEMICALS; COATINGS; EMULSIFIERS; SAFETY GLASS
POLYVINYL CHLORIDE-ACETATE COPOLYMERS	PHONOGRAPH RECORDS; FLOOR TILES; SOLUTION COATINGS; PVC PIPE
CELLULOSE ACETATE	FIBERS; CIGARETTE FILTERS; PLASTICS
ACETYL SALICYLIC ACID	ASPIRIN
ACETATE ESTERS	SOLVENTS
CELLULOSE ETHERS	DETERGENT PROMOTERS; TEXTILE CHEMICALS
	HERBICIDES
	ALKYD RESINS; SYNTHETIC LUBRICANTS
	BLEACHING AGENT; CATALYST
POLYVINYL CHLORIDE	PIPE & TUBING; SHEET; CONSTRUCTION MATERIALS; TEXTILE & PAPER COATINGS; FILM; INSULATION & PACKAGING MATERIALS
TRICHLOROETHYLENE	METAL DEGREASING SOLVENT
PERCHLOROETHYLENE	DRY CLEANING & METAL CLEANING SOLVENT
METHYL CHLOROFORM	METAL DEGREASING & CLEANING SOLVENT
POLYVINYLIDENE CHLORIDE & COPOLYMERS	FILM; COATINGS FOR FOOD PACKAGING
	CARBAMATE FUNGICIDES; CHELATING AGENT
TETRA ETHYL LEAD	GASOLINE ADDITIVE
	REFRIGERANT; SOLVENT; ETHYLATION AGENT
	FLAME RETARDANT FOR FIBERS & PLASTICS
	LEAD SCAVENGING AGENT
	ETHYLATION AGENT; SOLVENT; REFRIGERANT
POLYSTYRENE-GP, HI, EXPANDED	PACKAGING; HOUSEWARE; APPLIANCES; TV CABINETS; FURNITURE; LIGHTING; CONSTRUCTION; TOYS
ACRYLONITRILE-BUTADIENE-STYRENE COPOLYMERS	PIPE & FITTINGS; APPLIANCES; AUTOMOTIVE USE; BUSINESS MACHINES; TELEPHONES; ELECTRICAL & ELECTRONIC APPLICATION
STYRENE-BUTADIENE LATICES	TEXTILE & PAPER COATINGS; EMULSION PAINT; ADHESIVE FOAM PRODUCTS
STYRENE-ACRYLONITRILE COPOLYMERS	APPLIANCES; AUTOMOTIVE USE; HOUSEWARE; FURNITURE; PACKAGING
STYRENE-BUTADIENE ELASTOMERS	TIRES; MECHANICAL RUBBER GOODS; FOOTWEAR, ADHESIVES; CHEWING GUM
UNSATURATED POLYESTERS	REINFORCED: BOATS, BATHTUBS, TANKS, PANELS, TRAYS NON-REINFORCED: DECORATIVE & ELECTRICAL APPLICATIONS
UNSATURATED POLYESTERS	COATING RESINS; ELECTRICAL INSULATION
	DYES; PHARMACEUTICALS; PESTICIDES
ETHYLENE-PROPYLENE TERPOLYMERS	AUTOMOTIVE PARTS; TIRES; HOSE & TUBING; WIRE & CABLE
	REACTION PROMOTORS, INITIATORS, CATALYSTS
	CORROSION INHIBITOR; RUBBER CHEMICALS; INSECTICIDES
SINGLE CELL PROTEIN-YEAST	ANIMAL FEED
	TOILETRIES; COSMETICS; INDUSTRIAL SOLVENTS & THINNERS; DETERGENTS; FLAVORS; DISINFECTANTS
	SEE ABOVE
ETHYL ACETATE	SOLVENT
	SOLVENT; MEDICINALS; PHARMACEUTICALS
	PLASTICIZERS; SURFACTANTS
	PLASTICIZERS; SURFACTANTS
	GRAIN PRESERVATIVES; CELLULOSE PLASTICS; HERBICIDES
	HERBICIDES; SOLVENT
	RIPING AGENT FOR FRUIT & VEGETABLES

derivatives and uses.

5. Storage, Shipping, and Handling

5.1. Storage of Ethylene

Pure ethylene—polymer-grade ethylene can be accepted as pure ethylene—has a normal boiling point of $-103.7°C$ and a liquid density of 0.57 kg/dm^3. At ambient conditions it is a superheated vapor of about the same density as atmospheric air. It is impractical to store commercial quantities of ethylene in the gaseous phase at moderate pressures, hence it is stored either in the liquid phase at reduced temperatures in specially designed above-ground tanks or in the vapor phase at high pressure and at a density equal to about one-half of the liquid density at the normal boiling point in underground salt domes or similar reservoirs specifically created for the purpose.

5.1.1. Liquid Storage under Pressure

Ethylene may be stored in the liquid phase at any temperature along its vapor pressure curve below the critical temperature [21, 22]. The critical temperature of ethylene is 9.20°C.

A relatively small volume of strictly operational storage is needed in ethylene producer plants. Such storage rarely provides for more than 4 h of containment in cylindrical tanks and at some convenient plant refrigeration level, often found to be −40°C and 1.33 MPa. In many instances a case can be made for additional pressure storage at this level to hold off-specification product and/or provide breathing capacity for downstream processing units. Such extra storage is normally provided in pressure spheres which have been installed in sizes up to 4000 m^3.

Figure 9 shows such a double sphere. The load of the inner sphere is directly transmitted over an equatorial girder into vertical support columns. The outer sphere creates a vapor-proof envelope around the inner sphere to hold the insulation but has no structural function. The spherical annulus between both may be evacuated to reduce heat losses. Tank pressure is usually maintained over a refrigeration system that is independent of the plant's own system.

The inner tank material must meet the impact test at the lowest possible temperature that the tank might become exposed to, taking into account the backup systems provided for pressure maintenance.

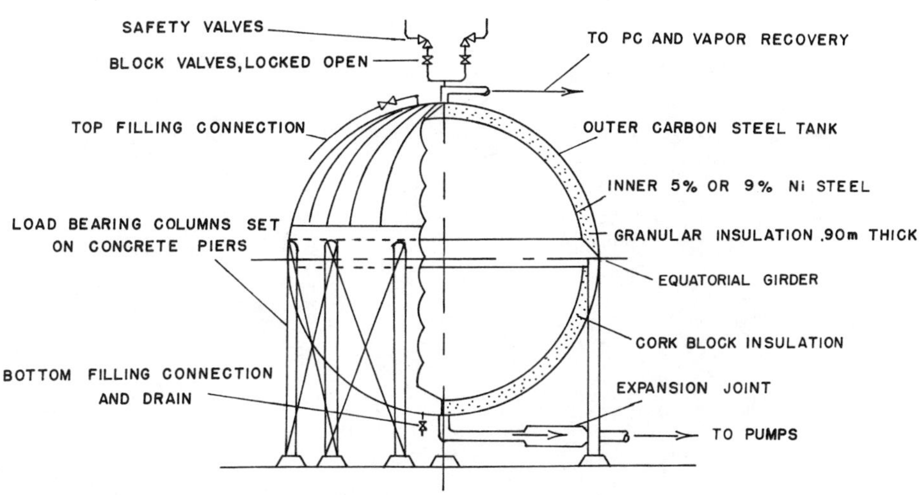

FIG. 9. Spherical ethylene storage tank.

5.1.2. Liquid Storage at Atmospheric Pressure

Large volume storage as a production reserve for downstream units and/or in seaside terminals for loading ethylene tankers is normally provided in aboveground, double wall cylindrical tanks having a practically square projection (height = diameter), at atmospheric pressure, i.e., with ethylene stored at its normal boiling point of −104°C.

Due to the high cost of such storage it is not usual to provide enough storage to assure the operation of downstream ethylene-consuming plants at their normal production rates during the periods of yearly maintenance of the ethylene plant. It has become normal practice in the operation of these plants and their downstream complimentary plants to do preventive maintenance work during scheduled intervals such that there ensues but a minimum loss in production and to judiciously reduce, during unforeseen difficulties in the ethylene producer, the feed rate to the downstream plants for the period that they might be depending on the storage inventory.

Plants that share in a pipeline distribution and storage grid may of course arrange for the continuous supply of ethylene to their downstream consuming units during the yearly shutdown of their producer plant.

Figure 10 shows an atmospheric pressure ethylene storage tank of early design. It consists of an inner alloy or aluminum tank fastened by hold-down bolts to a ringwall foundation, load bearing, foam glass insulating brick pad over compacted fill and an outer carbon steel tank similarly fastened. Heating coils in the tank base keep the compacted earth fill underneath from freezing to prevent frostheave damage to the tank floor. With this type of construction the hold-down bolts of the inner tank, however, largely prevent a radial contraction

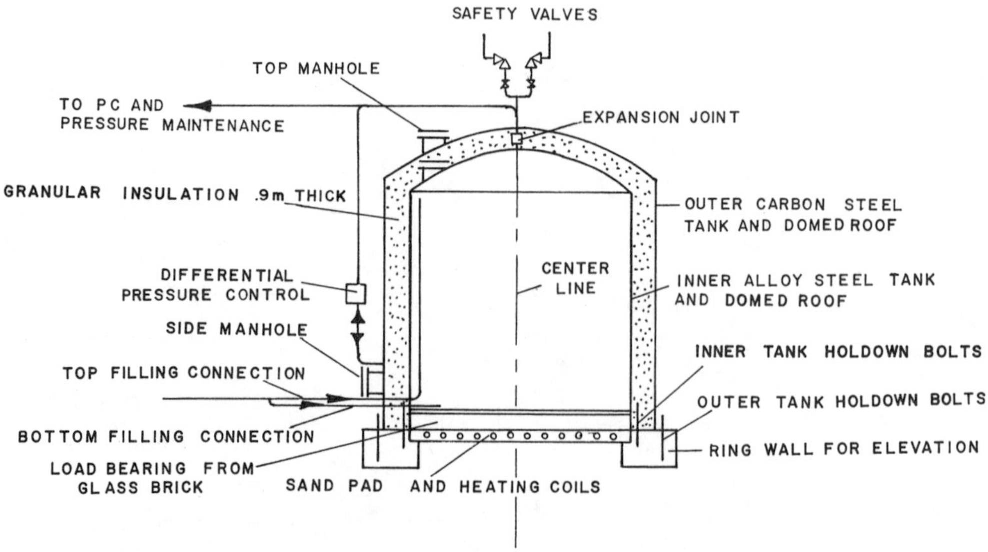

FIG. 10. Early design cylindrical storage tank for ethylene.

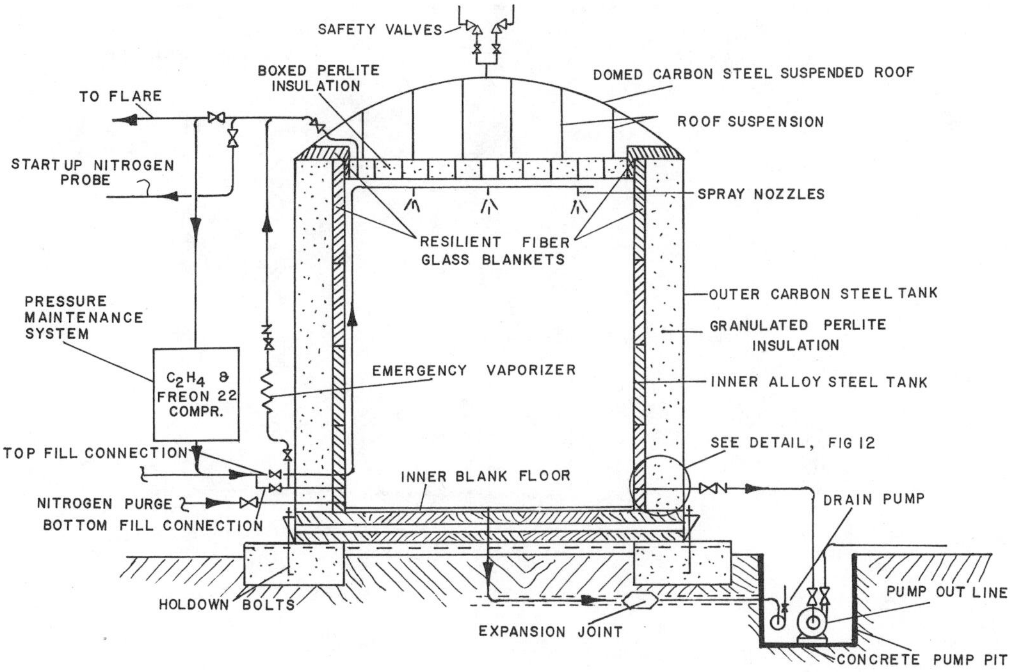

FIG. 11. Modern ethylene storage tank.

of the floor plates due to temperature changes, thus causing excessive stresses in the base angle area. Also, both inner and outer tank require separate pressure maintenance systems against changes in atmospheric pressure [23, 24].

Figures 11 and 12 show a late tank design that eliminates these objections by providing a suspended insulated roof over the inner tank with peripheral marginal space between roof rim and inner tank closed by flexible gas permeable foam glass blankets that permit the pressure to equalize between inner tank and tank annulus and disposes of the inaccessible inner tank hold-down bolts. The completed tank installation consists of the tank proper, ethylene transfer pumps, and a refrigeration unit serving the pressure maintenance system to handle the vapors which result from environmental heat input and from stock movements into and out of the tank. There is usually provided an emergency vaporizer which acts only in instances when the tank pressure is too rapidly depleted.

Inner tank contraction on filling is quite appreciable. In a 15,000-t ethylene tank, 32.8 m i.d., it will amount to 4.6 cm over the diameter if the inner tank is made of stainless steel and twice this amount if of aluminum. The outer tank may be made of carbon steel or of concrete, pretensioned vertically and circumferentially. The annulus between the inner and outer tanks is normally filled with granulated Perlite as insulation. To prevent the Perlite from dropping into the void space created by the tank's contraction, where it would impair reexpansion if the tank needed to be warmed up for inspection or repairs,

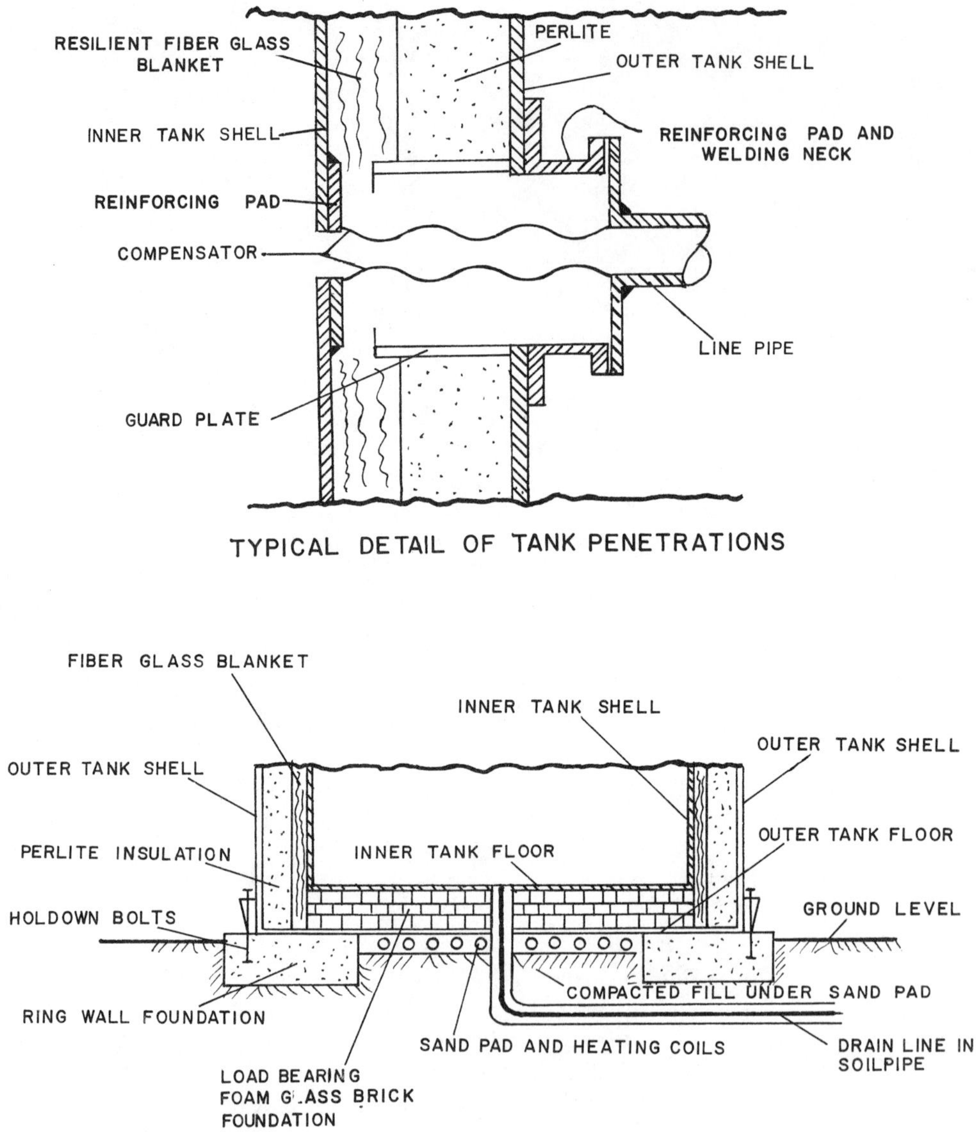

FIG. 12. Details of tank penetrations and tank floor.

fabricators suspend resilient, fiberglass blankets circumferentially around the outer face of the inner tank.

5.1.3. Safety in Tank Design, Technigaz Tank

The subject of safety in ethylene plants is briefly discussed in Section 7.4. Safety in the design of cryogenic storage tanks to store nitrogen, LNG, and ethylene is a matter of equal importance but somewhat peripheral to plant safety. The largest number of such tanks of double wall metal construction as described in the preceding paragraphs is in service in the over 100 peak load shaving stations scattered over the United States, Canada, and Europe to hold peak load LNG. Except for the material of the inner liner, which may be of a lesser grade due to the higher atmospheric boiling point of ethylene, tanks to hold ethylene or LNG are alike and present identical operational and environmental safety problems. Notable progress toward the ideal of an "ultimately" safe tank design has been made recently [201].

Technigaz in France developed a cryogenic tank design for the containment of large volumes of LNG and ethylene that deserves mention in this connection. The Technigaz tank (Fig. 13) comprises a reinforced and pretensioned outer concrete tank with a monolithic foundation as the secondary barrier, insulation mounted against the inner face, and a "waffle iron," preshaped, thin membrane liner as the primary barrier. Few of these tanks have as yet been installed in ethylene plants, due to their somewhat higher cost, but several design improvements make this design meet the trend of the times.

As particular safety features must be mentioned (1) that there is no movement between liner and foundation slab specifically at the lower base angle where stresses in a standard tank with flat base plates are usually highest, (2) that contraction and expansion are taken up differentially by the orthogonal corrugations of the inner tank membrane, and (3) that there are no peripheral penetrations. All lines enter and leave through the roof of the tank. Discharge is by means of deepwell pumps, so a lock and/or other safety measures have to be provided on the roof for the safe removal of a pump in need of repair. The Technigaz tank is marketed in the United States by Kelgas, a division of Pullman-Kellogg.

Progress also continues to be made in the development of high-strength insulating materials. Bayer A.G. in West Germany has developed a polyurethane foam, marketed under the trade name PUR foam, that possesses excellent properties and bids well to reduce the expense for insulation, a high cost item, of cryotank installations. The Dutch firm Dicta International is offering tank designs utilizing these materials in combination with a concrete outer tank [25].

5.1.4. Ethylene Tank Materials and Costs

Table 4 shows the tensile strength values of plate materials commonly used for ethylene tanks, according to API 620, Appendix Q.

In addition, 5.0 and $5\frac{1}{2}\%$ DIN standard Ni steels are now also accepted but require great care in fabrication as they are difficult to weld. Table 5 gives approximate (mid-1977) erected costs for a U.S. Gulf Coast location of ethylene storage tanks including tank fittings, ringwall foundation, and $5\frac{1}{2}\%$ Ni steel inner tank but exclusive of pumps, pressure maintenance system, and other appurtenances [26].

Codes and standards that apply to ethylene storage tanks in the United States in general are contained in publications of the following rule-making

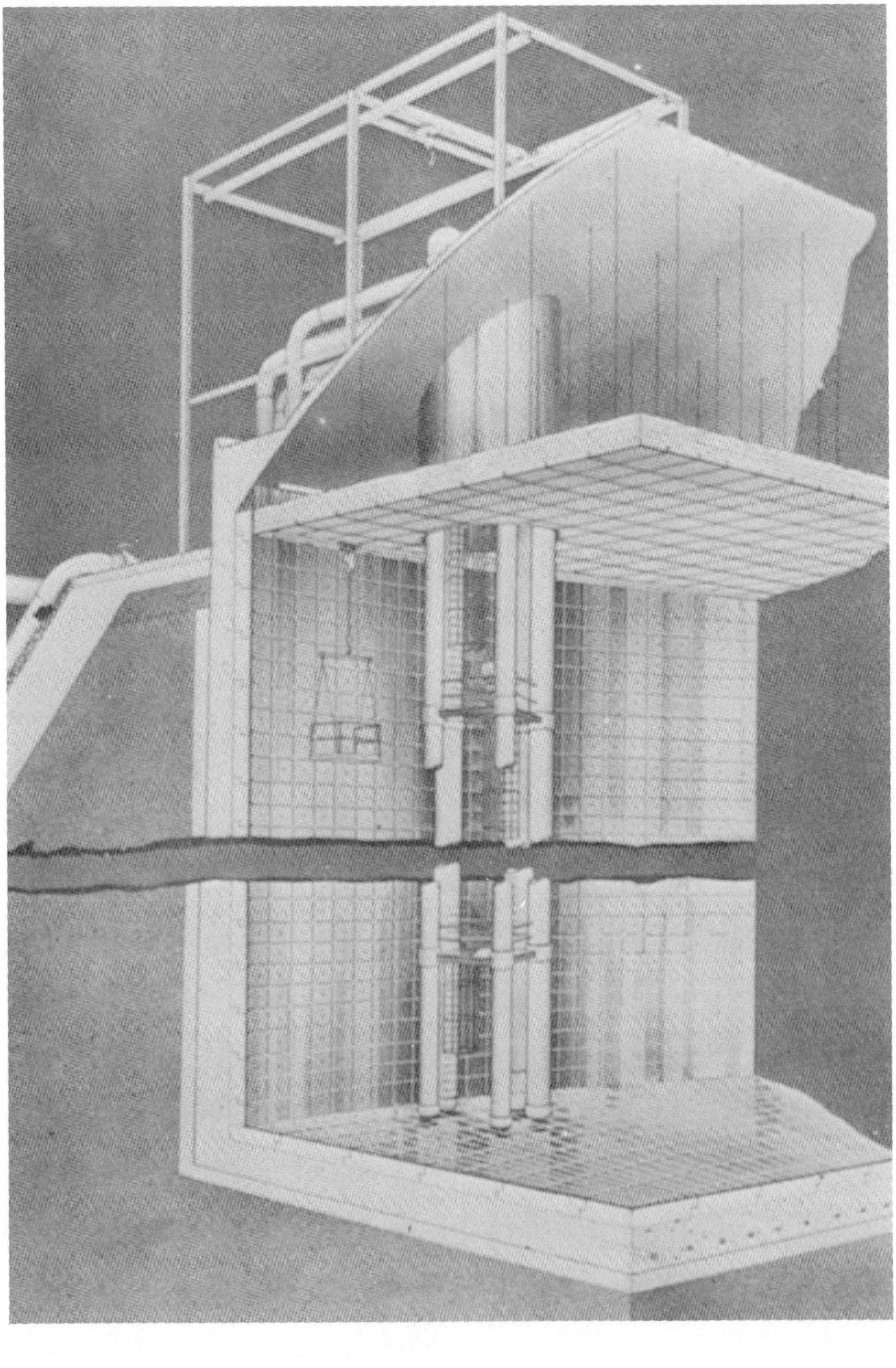

FIG. 13. Technigaz ethylene tank.

TABLE 4 Tank Materials

Material	ASTM No.	Allowable Stresses Design (MPa)	Allowable Stresses Test (MPa)
9% Nickel	A 353-64	197[a]	290[a]
9% Nickel	A 353-65	197[a]	290[a]
5% Nickel	A 645-74a	218	290[a]
Stainless steel	A 240-304	138	165
Aluminum	B 209-5083-0	82.7	99.3

[a]Based on strength of deposited weld metal.

bodies: NFPA (National Fire Protection Association), API (American Petroleum Institute), ASTM (American Society for Testing and Materials), and ASME (American Society of Mechanical Engineers).

The development of ultimately safe cryogenic storage tanks cannot be said to be complete at this time. The latest designs feature an external concrete bund (instead of a low height dike) around the tank and provide for all tank penetrations in the tank roof, as noted above.

5.1.5. Tank Testing and Commissioning

Testing of an above-ground atmospheric pressure ethylene tank prior to placing it in service involves procedures that have been worked out in detail by tank fabricators everywhere and are identical with those for the testing of LNG tanks. In the case of double wall tanks, the inner tank is normally tested for tightness by a vacuum box technique before the fiber glass expansion blankets are hung from the inner tank. In the Technigaz tank tightness is tested by painting the seams with a color sensitive dye solution and applying air pressure to the reverse side.

The tank is readied for the hydrostatic test after all control and safety valves have been installed and operating lines have been blind-flanged. The test is carried out with water, preferably of low chlorine content, to a level equal to 1.1

TABLE 5 Approximate Erected Tank Costs [27]

Nominal Volume (m^3)	Cost (million $)	Unit Cost (U.S. $/$m^3$)
6,360	1.14	179
12,720	1.73	136
19,080	2.28	119
23,800	2.65	111

and up to 1.25 times the maximum operating hydraulic head. As the tank is being filled and for a 15 to 30 day period afterwards, settlement of the ringwall foundation is carefully determined with a transit instrument. Differential settlement of the central part of the foundation may be ascertained by the use of a plumb line through a roof nozzle. At the end of the period, the tank design pressure is produced with air to set all pressure regulators and safety valves to previously certified values. Should a tilt beyond admissible tolerances develop during the hydrostatic test, the tank foundation must be releveled.

Normally, after draining and mopping up the inner tank, manhole covers are welded shut, the granular Perlite insulation is blown in, and the tank is vibrated with suitable tools to compact the insulation to the desired degree. Finally the tank is purged of air with relatively pure nitrogen. If done properly, two tank volumes of nitrogen suffice to reduce the residual O_2 content to where it will not adversely affect the purity specification when the tank is filled with liquid ethylene.

5.1.6. Storage in Salt Domes

The rapidly growing demand for ethylene over the past 30 years has led to the construction of ethylene pipeline systems that connect producers with consumers. Such systems exist in the United States, Western Europe, and to a very limited extent in the Comecon countries of Eastern Europe. All these systems transport ethylene in the vapor phase. For their smooth and purposeful functioning, the development of large, high-pressure, leak-proof underground reservoirs to hold an adequate production reserve was a prime necessity. Massive salt formations underground proved ideal for this purpose.

Rock-capped salt dome structures, suitable for the storage of petroleum products such as LPG, have been located in many parts of the United States, but are particularly numerous, onshore and offshore, along the U.S. Gulf Coast. In 1975 some 22 domes in Texas and Louisiana with a capacity of close to 16 million m^3 were engaged in LPG operations [28]. Still others are in use to store ethane, NGL, and ethylene, and work is proceeding to enlarge natural gas storage capacity at very high pressures [29]. Similar ethylene reservoirs are in operation in France at Viviat as part of the pipeline system connecting Feyzin with consumers at Travoux (Solvay) and Pont de Claix (Progil), and connecting Lavera (Naphtachimie) with Berre (Shell) and St. Aubin (Pechiney-St. Gobain) in the Provence.

Reservoirs are also planned for the ARG grid (Aethylen-Rohrleitungs-GmbH), one of the large European distribution systems, connecting producers and consumers in Germany, Belgium, and Holland.

5.1.6.1. Salt Dome Storage—Creation and Operation. Nearly all of the dome cavities were leached specifically for storage; a few were created during the process of obtaining brine for chemical plants. As an example, Fig. 14 shows the (1955) Fannett ethylene reservoir of Gulf Oil near Port Arthur, Texas. Gulf Oil did considerable pioneering work in the development of ethylene reservoirs. The Fannett dome contains four storage wells: two for ethylene, one for

DEPTH m	PRESSURE IN BRINE STRING MPa	PRESSURE IN ETHYLENE STRING MPa	LOCAL TEMPERATURE °C
0		9.24	20.
152.	1.79	9.71	29.
305.	3.59	10.2	37.
457.	5.38	10.6	41.
610.	7.17	11.0	43.
762.	8.96	11.4	45.
884.			
914.	10.8	11.9	47.
1036.	12.2	12.2	48.
1189.			

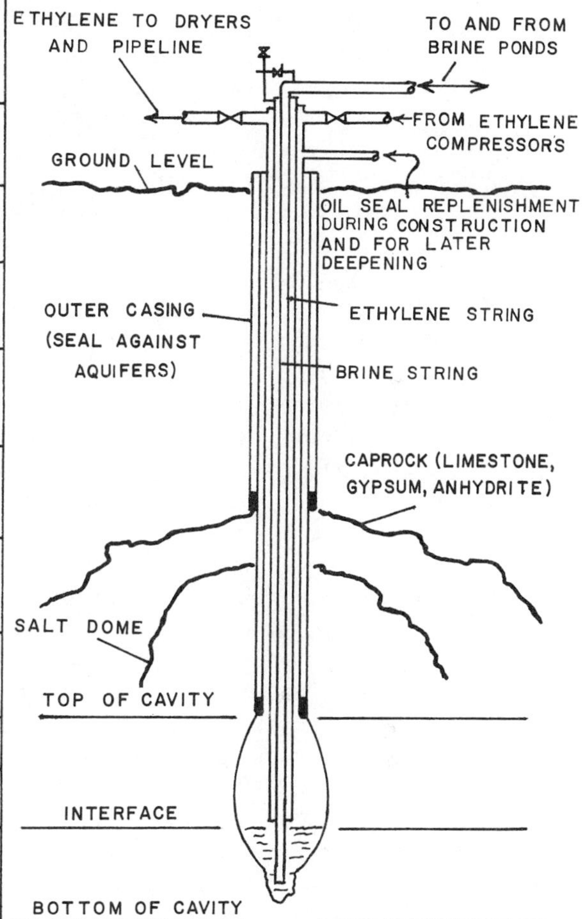

FIG. 14. Fannet salt dome reservoir (schematic).

propane, and one for butane. A prerequisite to the construction of a salt dome cavity is a plentiful source of water for the leaching operation and a means of disposing of the brine without harm to the environment. A well is drilled through the caprock as in drilling for oil and the necessary strings of casing are cemented in place. The mining or leaching operation can be done from the top down with movable tubing. The preferred procedure is to leach from the bottom up with a layer of oil (guard fluid) floating on top of the leaching agent (fresh or brackish water) to have some control over the shape of the cavity.

At Fannett, for instance, caprock was reached at about 650 m and drilling continued to the 1190 m level. The leaching operation started at this level through an annular leaching string, the brine returning to the surface through a central dip leg. While leaching goes on, careful measurements of the flow and salt content of the effluent are taken to obtain a continuous check of the volume of cavity created. An average figure is from 8 to 10 m^3 of brine circulated per cubic meter of cavity created. The result of the leaching operation is a spindle-shaped void. Fannett at the time of completion (1957) had an estimated total volume of over 65,000 m^3. The dimensions of the cavity were given as 300 m in height by 20 m maximum diameter [30]. Normally the depth of the cavity varies between 900 and 1500 m. Fannett, while among the smaller reservoirs serving the Texas ethylene grid, is typical of similar installations elsewhere [29, 31, 32].

The design is dictated by the behavior of ethylene around the critical region (5042 MPa and 9.2°C). Ethylene throughout the cavity has dense phase properties since it is stored above a column of brine which at the underground interphase exerts a pressure much in excess of the critical pressure of ethylene.

To enter the cavity, the ethylene must be compressed to a pressure greater than the difference between the weights of the brine and ethylene columns, and the depth of the cavity must be such that the pressure is at all times greater than the critical pressure. On its way down to the cavity, the temperature of the dry ethylene, which may have been recooled to 25°C after compression, will increase due to the increase with depth of the temperature of the surrounding substrata. In Texas and Louisiana the bottom temperature of salt dome cavities is found to be between 40 and over 50°C, well above the critical temperature of ethylene (9.2°C), so that subcooled liquid ethylene and a vapor interphase cannot exist at these conditions.

As ethylene is pressured into the cavity, brine is being displaced into lined surface ponds, and while ethylene is being withdrawn, brine by gravity or low head pump is being returned to the brine column. At the interphase level the weight of the brine column balances the weight of the dense phase ethylene column plus the external pressure superimposed on it (see Fig. 14). As ethylene is being withdrawn, the pressure in the cavity diminishes and the level in the brine string descends below grade even as brine is being returned. The pressure reduction produces a cooling effect in the ethylene leg and a gradual increase of the water content. If ethylene were discharged from the cavity at an uncontrolled high rate, the effect would cause stratification and an erratic discharge rate due to swings in ethylene density. The plot of Fig. 15 of density vs temperature with pressure as the parameter shows the pronounced change of density with pressure in the region above the critical pressure and operating points for various salt domes.

At ground level the ethylene is released into a knockout drum from which water may be withdrawn if the temperature is maintained above the hydrate inception point (20°C). It is here and in the upper portion of the ethylene tubing that hydrates may form if the withdrawal rate is too high or the temperature too close to the hydrate point.

From the knockout drum the ethylene is passed over dryers and into the pipeline. To determine mass flow during discharge requires the measurement of volume flow, pressure, and temperature. The three measurements may be combined over a computing relay and the result fed to an integrating flowmeter or else the density may be measured directly and combined with the flowmeter reading to obtain the mass flow.

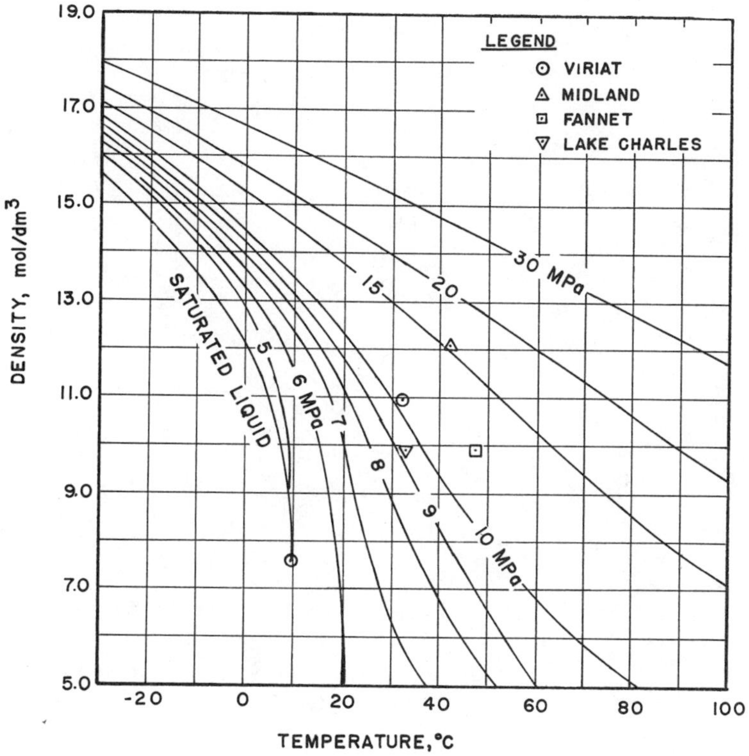

FIG. 15. Density diagram for ethylene showing operating points for various salt domes.

5.1.6.2. Storage in Multilayered Salt Strata Caverns. In different parts of the United States and abroad, salt occurs in stratified layers from 5 to ~25 m in depth that are separated by layers of limestone, dolomite, and anhydrite. In the absence of suitable salt domes these layered salt deposits may be leached by essentially the same technique described earlier to create cavernlike storage voids in a vertically tiered arrangement. By proper logging and tooling during the drilling operation, it is possible to connect caverns at different levels (Fig. 16).

The Dow Chemical Co. in the late 1950s near Midland, Michigan, developed the concept of cavern storage and has successfully operated an ethylene cavern storage system since that time [33].

The principal differences of the cavern type compared with a salt dome cavity are the dimensions and shape of the void. The latter extends mainly in a vertical, the former in a horizontal direction. In the creation of a cavern system it is essential to be assured of the nonpermeability to ethylene of the surrounding rock strata and of the soundness and security of the caprock roof and in-between layers of rock against cave-ins that could damage the tubing. From the operational point of view, both types are similar except possibly for the type of brine that is used to displace the ethylene on withdrawal. If the rock strata exposed in the cavern are rich in calcium and magnesium, calcium sulfate

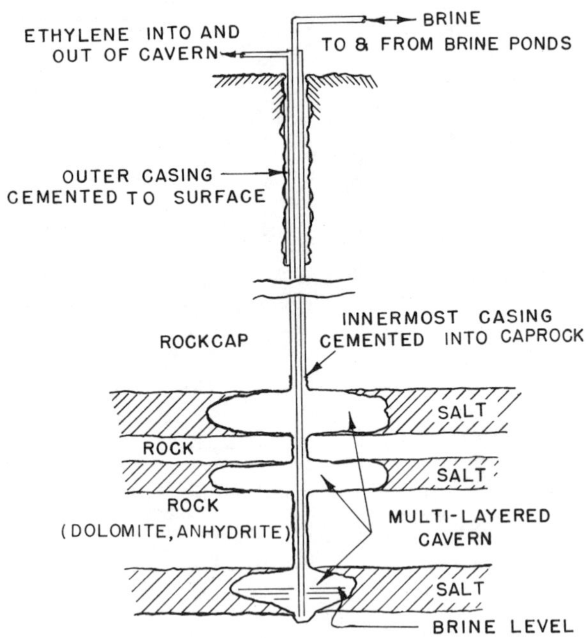

FIG. 16. Ethylene storage in multilayered caverns (Dow Chemical Co.).

crystals (gypsum) will precipitate in the brine and will tend to clog the brine string. At Dow-Midland this problem was solved after completion of the cavern by using a substitute brine for displacement [33, 34].

5.2. Transportation of Ethylene

During the 40+ years that ethylene has become the most important hydrocarbon for chemical synthesis, additional means had to be devised to transport and store it in ever-increasing volume.

5.2.1. Transport in Cylinders

Before that time ethylene was moved in steel cylinders either singly or in an assembly of cylinders mounted on a truck platform. The cylinders were designed for a maximum working pressure of 11.8 to 13.7 MPa gauge and a maximum temperature of 45°C. At these conditions, ethylene is a dense phase gas of about 300 kg/m^3, about 250 times its density value at ambient conditions. A cylinder of customary size would have a volume from 55 to 70 dm^3. Designed in accordance with material specifications of the day, the tare of a 55-dm^3 cylinder would be about 80 kg and it would contain $0.055 \times 300 = 16.5$ kg of ethylene, so that nearly five times as much metal had to be handled per unit of product. Commercially insignificant amounts are still so distributed today.

5.2.2. Transport by Tank Truck

In 1951, ICI in England, after careful preparations, placed into operation a fleet of tank trucks to haul liquid ethylene at atmospheric pressure from their Billingham plant to another site 225 km or about 6 road hours distant. When simulating road operations over an extended 8-d period it was found (1) that the tank pressure rose from 60 kPa gauge to 735 kPa gauge while the safety valves were set at 1040 kPa gauge, and (2) that on standing it rose considerably faster due to nonequilibrium conditions existing at the liquid–vapor interface. The empty vehicle weighed 1.9 t/t of payload.

Subsequently similar vehicles were developed as truck-trailer units in the United States and elsewhere, and the tank insulation was made more effective. A truck-trailer unit of typical design [35] today has a light weight of about 15 t and can carry a maximum payload of 20 t of ethylene. Maximum tank operating pressure is held to about 324 kPa gauge. By operating a fleet of these units, Esso-Imperial of Canada during the 1960s supplied two ethylene consumer plants in the Eastern United States with their ethylene requirements on a regular basis from Imperial's Sarnia, Ontario, ethylene plant. Truck-trailer units have also proven their worth in Europe during start-up operations of newly constructed plants and during lengthy shutdowns of existing plants due to accidental failure or fire damage.

5.2.3. Transport by Railroad

Simultaneously with road tankers, railroad tankcars were introduced that met all applicable safety standards. The standard ethylene railroad tank car manufactured by Lox Equipment Co. of Livermore, California, one of the prominent United States car builders, has a light weight of 52.6 t and carries a maximum payload of 65.3 t of ethylene. The annular space between an outer carbon steel and an inner type 304 steel tank is filled with expanded Perlite and evacuated to minimize heat losses. While single rail tankers are not competitive with road tankers over relatively short distances because of their slow speed of movement, they might over longer distances of up to ~ 1200 km be competitive with road tankers if the volume were adequate to be moved in unit trains. A unit train of 20 tank cars could move 36,500 t/yr over a distance of 1200 km. Yet, the incremental cost of movement by road or rail may burden the cost of the ethylene derivatives plant to make the enterprise nonviable.

A number of ethylene rail tankers are moved by American railroads under circumstances where time is not of the essence. One example involves a movement of ethylene over a distance of about 5150 km. The LNG plant of Phillips Petroleum at Kenai, Alaska, is operated with a three fluid refrigeration cascade. Make-up ethylene is supplied to it by an ethylene plant at Odessa,

Texas, by rail to Seattle, from there by ferry to Whittier, Alaska, and by rail to Crown Point, Alaska, where it is transferred to a truck-trailer unit for the last 110 km trip to Kenai. Transit time averages 18 to 20 d. The tank annulus is highly evacuated and the car is routed to avoid layovers at transfer points, so that tank pressure increases to only 380 kPa gauge at Crown Point [35].

5.2.4. Transport by Oceangoing Tanker [36, 37]

Spotloads of ethylene have on several occasions been moved across the oceans, but regular intercontinental traffic in ethylene does not yet exist. In mid-1974 about two dozen tankers, most of them of a payload capacity under 2000 t, were engaged in coastwise traffic in European waters, the Carribean and between the islands of Japan. Lately two ships with four Kvarner-Moss spherical tanks each have been commissioned, and others are building, that can load alternately LNG, ethylene, LPG, or ammonia, the ethylene capacity being 16,500 t at −104°C [38, 39]. As LNG carriers these tankers can profitably be employed only over short distances and/or for modest import volumes. As ethylene carriers they will have an intercontinental range, limited only by the storage available at ethylene import terminals. There exist presently only two such terminals at which a 16,500-t ethylene tanker could discharge. To lower the ethylene transport cost, very large capacity combination tankers (125,000 m^3 and more) that would carry a proportionate amount of ethylene with a base load of LNG in separate tanks have been proposed [40].

Ethylene was first moved in 1965 by the *No. 1 Ethylene Maru* in Japan. She carries 108 t of ethylene in a cylindrical tank at −30°C and 1.81 MPa gauge pressure. Had the tank been designed for atmospheric pressure, the capacity could have been increased by 30% which points to the dominant effect of temperature on density. It is of importance to operate an ethylene tanker fully refrigerated to make full use of its hull dimensions; in addition, it was recognized early that pressure tankers can discharge readily only into pressurized shore storage which is always of limited capacity. Thus with the exception of some four early Japanese-flag carriers of small capacity, all tankers now in service provide for load containment at atmospheric pressure [41, 42].

Ethylene carriers, like LNG, LPG, and crude oil carriers, return to their loading ports in ballast. Crude oil tankers carry their ballast water in a number of oil tanks according to a predetermined ballasting plan that guarantees maximum stability and seaworthiness of the tanker on the empty voyage. With ethylene and LNG carriers, ballast capacity is provided in separate tanks between the outer hull and cargo tanks and in the double bottom.

Ethylene carriers mount a refrigeration unit for reliquefaction of the boil-off during transit and while in port since ethylene is too valuable to vent. Refrigeration is vital to the carrier's operation, hence full sparing must be provided. Reliquefaction units normally comprise an oil-free, labyrinth sealed ethylene piston compressor and a two-stage Freon 22 compressor, both motor driven, plus the necessary number of heat exchangers and condensers. There are also installed on board an inert gas generator and load vaporizer as well as a high-capacity, low-head vapor compressor for the return of displacement vapors while loading.

Operation on the high seas of tankers is under the nominal supervision of IMCO, the Intergovernmental Maritime Consultative Organization with 14 member countries. IMCO, in cooperation with the International Chamber of Shipping (ICS) and the International Association of Classification Societies (IACS), after years of labor has recently issued an all-inclusive code "to provide internationally agreed standards on design, construction and operation of vessels for the safe carriage of liquefied gases in bulk" that is now up to the member countries for ratification. IMCO has no powers to take corrective or punitive action against code violators which can only be initiated through the country of registry. However, IMCO member nations will be obliged to deny entry into any of their ports to violators. Thus IMCO will provide a powerful forum for the advancement of the art of constructing and operating ships collectively known as liquefied gas tankers, to which ethylene carriers belong as one class [43].

5.2.5. The Future of Marine Transport

There were in operation in 1975 throughout Western Europe 15 ethylene terminals with a combined atmospheric storage capacity estimated at about 40,000 t [44]. Intracontinental traffic between these points in 1975 amounted to about 300,000 t of ethylene [41]. Much of this traffic was between company-owned terminals, such as Teeside-Rozenburg (ICI), Terneuzen-Stade and Terneuzen-Tarragona (Dow), and Brindisi-Porto Maghera (Montedison). Some 7.5 million tons of ethylene were produced in Western Europe in 1975, so that maritime shipments amounted to only 4% of production.

Over the next decade this situation will change gradually. As the North Sea fields come into production, it is expected that Britain will achieve independence of crude imports and might even have available a surplus in distillates suitable for olefins production by 1985. But more important will be the change wrought by the OPEC countries becoming olefins producers over this period.

Particularly the Arab countries Iraq, Saudi Arabia, Qatar, Abu Dhabi, Algeria, and Iran will have available huge amounts of high yield feedstocks such as ethane and NGL fractions that they will wish to utilize for the production of olefins and downstream products.

Earlier studies have shown [45] that the cost of ocean transport depends on two parameters; (1) the one-way distance and (2) the capacity of the tanker. The break-even transport cost t may be calculated from the linear relation

$$t = (a + b)d$$

d being the one-way distance in kilometers and the constants a and b being functionally related to the capacity of the ship. To obtain the total transport cost, capital and operating charges for the terminals, which will be substantial and which vary directly with the ship's capacity, must be added. The data of

TABLE 6 Cost of Ocean Transport of Ethylene

Type No.	Nominal Tanker Capacity (t)	U.S. $/t, One-Way Distance (km)		Refs.
		5,550	11,000	
1	2,240	102	188	26
2	6,700	56	97	26
3	16,700	31	56	26
4	27,800	20	36	27

Sike [46] confirms the above proportionality relationship. Table 6 contains data on the transport of ethylene for two distances according to Sike.

Type 1 refers to a coastal tanker without viable capacity for intercontinental distances. At the upper end of the capacity scale, one Type 4 tanker [51,000 m^3, 35 km/h (19 knots)] could trans-ship the output of a large and distant ethylene plant as shown in Table 7. For distances under 9000 km the capacity of this tanker cannot be utilized economically because single loadings from one port will never reach a volume equal to the delivery capability of this tanker over a distance of 9000 km.

At present the movement of ethylene by coastal tankers is greatly inferior to the volume handled by the most important conveyance of all, the ethylene pipeline, but it is reasonable to expect an about five- to sixfold increase in movements by intercontinental tankers in parcels of 7000 tons to a maximum of

TABLE 7 Distances and Ethylene Deliveries [one 51,000 m^3 ship, cruising speed 35 km/h (19 knots)]

From Dammam, Saudi Arabia to	One Way (km)	Round Trip Time (d)[a]	Trips per Calendar Year[b]	Deliveries (t/yr)
Tokyo, Japan	12,200	34	9.85	283,000
Santos, Brazil	16,500	44	7.60	218,000
Buenos Aires, Argentina	16,900	45	7.45	214,000
Rotterdam (via Suez)	12,000	33.5	10.0	287,000
Sidney, Australia	13,500	37.0	9.05	260,000
From Skikda, Algeria to				
Philadelphia, Pennsylvania	7,500	23	14.6	418,000
Rotterdam, Holland	3,780	13.9	24.1	693,000(!)

[a]Includes foul weather allowance, 0.75 d, and terminal layover (pilot to pilot), 1.75 d/one-way trip.
[b]335 operating days, 30 d dry-dock allowance per calendar year.

about 20,000 tons over the next 8 to 10 years. The recent commissioning of two 30,000 m^3 LNG/ethylene/LPG tankers (16,200 t of ethylene) appears to have been motivated by prospects in the olefin feedstock and ethylene product transportation situation.

5.2.6. Transport by Pipeline

5.2.6.1. Historical Development and Present Status. A few years prior to and during World War II the demand for ethylene increased drastically. By 1950 a cluster of ethylene producers, whose work was based on the pyrolysis of hydrocarbons, and of ethylene consumers had come into operation in the Texas City–Houston–Lake Charles area of the United States to lend support to the idea of pipeline connections for the benefit of producers and consumers alike.

As ethylene production expanded, so did the grid of ethylene product lines, as the umbilical cord between producers and consumers and through the inclusion of storage cavities provide for a substantial production reserve [38]. In Texas (1974) there were 22 producer-consumers and 10 consumer companies connected by 10 proprietary lines of up to 250 mm diameter and backed up by nine salt dome reservoirs of a total volume reported to exceed 3 million cubic meters. Since that time the grid has been further enlarged to where (early 1978) it now has access to 51 manufacturing, consumer, and storage installations as shown in Fig. 17. A similar but smaller grid that counts four producer-consumers, three consumers, and two storage domes grew up in Louisiana. The Texas and Louisiana grids were recently connected by a 200-mm Dow line (see Fig. 18).

In 1962 a 150-mm diameter, 125-km long, 72,000-t/yr ethylene line connecting a producer at Esso's Fawley refinery with ICI's Severnside ethylene oxide and glycol plant was constructed in England [47]. Other lines in England and on the Continent followed in quick succession.

The 1974 status of the European ethylene grid is shown in Fig. 19 [48]. This grid had its beginnings in local grids in the Cologne–Ruhr and Antwerp–Rotterdam areas which were eventually joined by the ARG (Aethylen Rohrleitungs Gesellschaft) line after a tie-in [49] with the Ludwigshafen–Kelsterbach–Cologne line had also been effected. Two one-way traffic (producer to consumer) lines in the South of France, the 278-km Etel-line (1967) out of Feyzin and the 125-km Lavera–Berre line (1969) supplying a PVC complex at St. Aubain were also connected [50].

In Britain the ICI road tanker service out of Billingham-Wilton (now Cleveland) was replaced early in 1968 by a pipeline [51, 52]. A new 250 km line from Wilton to Grangemouth (BP) is under construction, and there might be further extensions when North Sea gas comes ashore further north. Wilton ethylene has been ferried from Teeside to Rozenburg, Holland, by the two 820 m^3 coastal tankers, M.S. *Teviot* and M. S. *Traquair*, since September 1966. Thus ICI theoretically could deliver ethylene into the West-European grid.

There are other ethylene pipeline transmission systems of lesser extent about to be commissioned and in the planning stage. Noteworthy among these

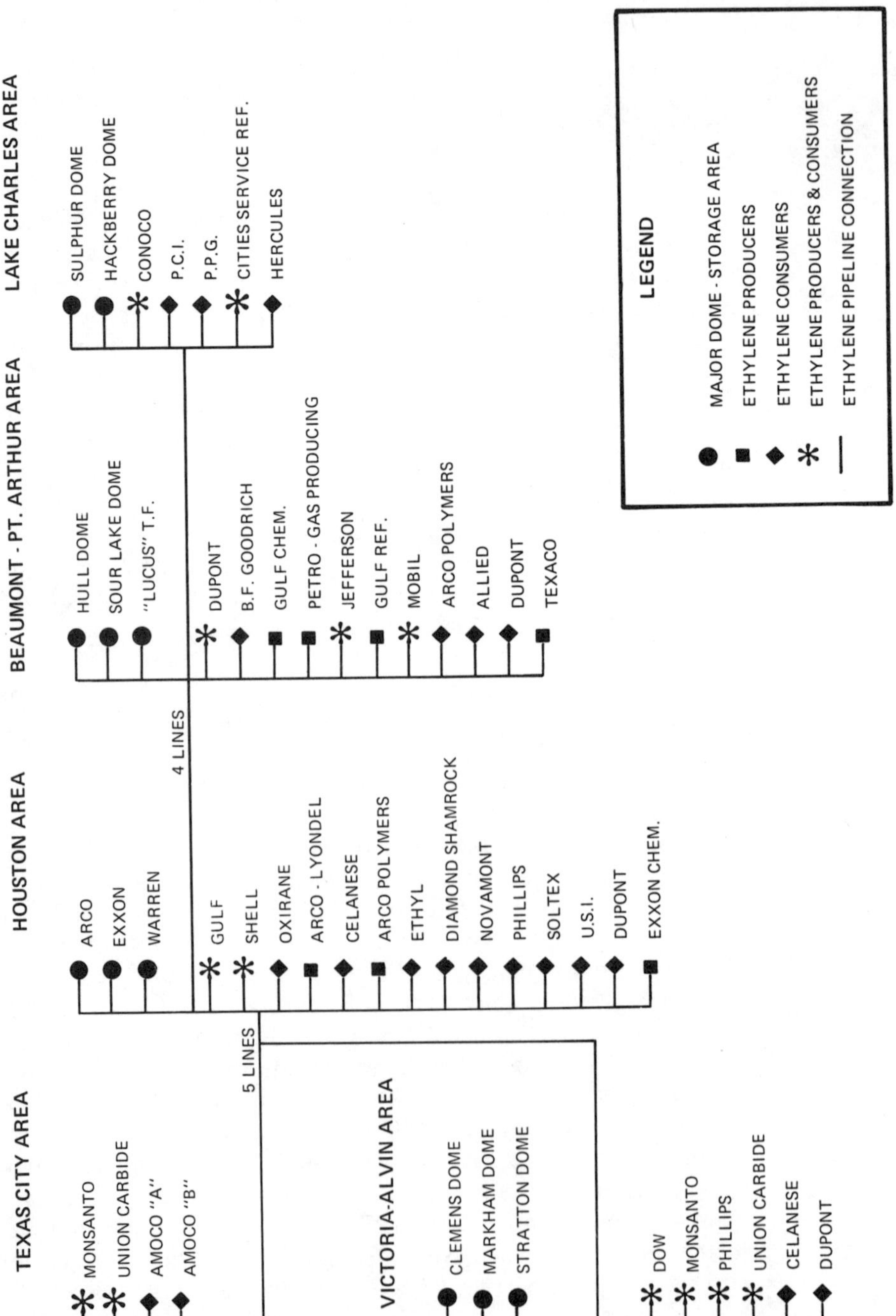

FIG. 17. Ethylene grid for greater Houston area. (Courtesy Dewitt Graphics, Houston, Texas.)

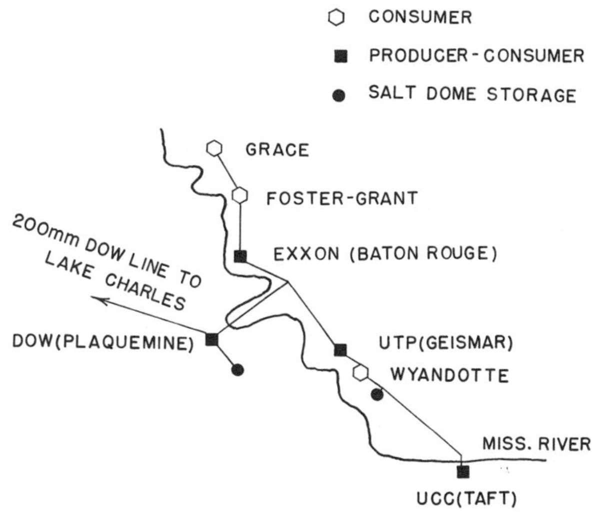

FIG. 18. Louisiana ethylene grid.

is a 90-km line from Tarragona, Spain, connecting the Enpetrol producer with a Solvay subsidiary consumer at Matorell, Spain. In the USSR two lines have recently been commissioned. A line in the Tatar and Bashkir Republics designed by Phillips Petroleum links five petrochemical centers containing three producers, two of them in operation and one in its late stage of construction. Another line connects the new Tisca chemical complex at Leninvaros, Hungary, with manufacturing facilities at Kalusch in the Ukrainian Soviet Republic. According to one report [53], a grandiose scheme to link 19 producers and 66 consumers in eight Comecon countries by a giant network of ethylene pipelines is to become a reality by 1990.

5.2.6.2. Ethylene Pipeline Design. The same fundamental considerations govern the design and operation of an ethylene pipeline as must be observed when designing a salt dome cavity or multilayered salt cavern. Two-phase existence in the pipeline must be avoided.

Refrigerated liquid lines for the transmission in one direction of very large volumes of LNG have been proposed. A 1090-km liquid, nonrefrigerated ethane line from Bushton (near Wichita), Kansas, into the Chicago area is operated as a common carrier and includes the Chemplex and Morris, Illinois, plants of Northern Petrochemicals. It has been in service since 1972 [54].

Conditions are different for an ethylene pipeline network because ambient seasonal ground temperatures straddle the critical temperature of ethylene. Compared with natural gas lines, distances, mass flow, and line sizes in an ethylene grid are small, and flow may be in either direction and highly variable.

It follows that for economical transportation by pipeline, ethylene must be transmitted in the vapor phase and at either subcritical or preferably supercritical pressure.

Figure 20 shows an envelope or design range of operating conditions of

FIG. 19. Ethylene pipelines in Western Europe. (Source: W. van der Linde, *Oel-Zeitschrift fuer die Mineraloelwirtschaft*, February 1975.)

several ethylene transmission lines in a temperature–enthalpy diagram. In EEC countries the maximum working pressure is set by law at 9.8 MPa. The range of conditions is given in Table 8. The ARG system pressure seems to slightly exceed the legal limit.

Due to the five- to eightfold difference in density in the subcritical versus the supercritical range, the transport capacity of a supercritical line is several times that of a subcritical line for the same pressure gradient. Van der Linde of the

ARG ethylene pipeline system has published an instructive diagram [49] that shows this relationship graphically, it is redrawn in Fig. 21. The graph is drawn for the specific case of a 250-mm line, 100 km in length, ethylene temperature 10°C. Isothermal flow conditions were assumed which presupposes that the cooling effect due to expansion is continually being compensated by heat transfer from the ground into the line. As an example, in the subcritical area maximum line capacity for a pressure drop of 5.0 MPa is shown as 400,000 t/yr of ethylene with an initial pressure close to 5.1 MPa. Obviously the entire available pressure gradient will be consumed over the 100-km line, i.e., the line exit pressure will be 0.1 MPa. Operating in the supercritical range by boosting the initial pressure to 6.1 MPa, the line pressure drop would be only 1.0 MPa and the line terminal pressure would be near 5.1 MPa for the same annual throughput of 400,000 t/yr.

While these results are idealized and leave no safety factor (the terminal pressure in the supercritical case dropping below and the initial pressure in the subcritical case exceeding the respective values listed in Table 8), the example

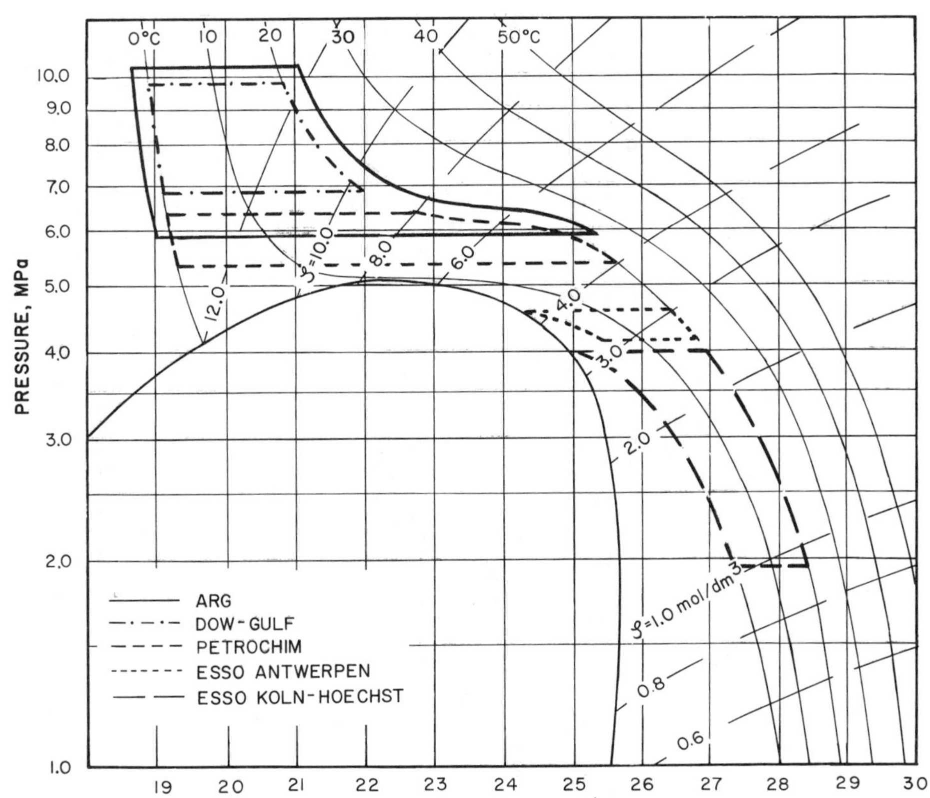

FIG. 20. Operating range for various ethylene transmission lines. (Source: W. van der Linde, *Oel-Zeitschrift fuer die Mineraloelwirtschaft*, February 1975.)

TABLE 8 Range of Pipeline Operating Conditions

Mode	Pressure (MPa)	Temperature (°C)	Specific Volume (m^3/kg)	Density (kg/m^3)
Subcritical	2.0 to 4.4	0 to 20	0.1 to 0.38	100 to 27
Supercritical	5.4 to 9.8	2 to 20	0.0025 to 0.005	400 to 200

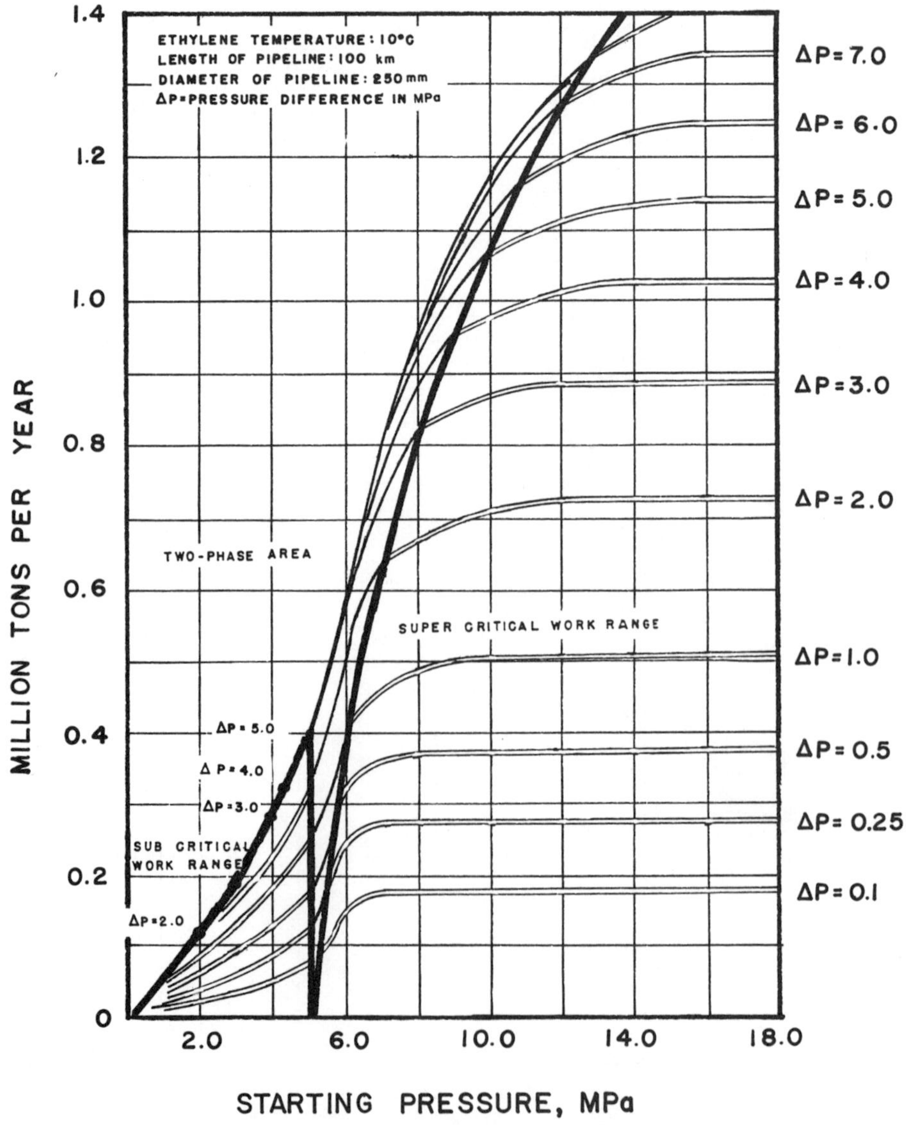

FIG. 21. Starting pressure vs line capacity (Source: W. van Linde, *Oel-Zeitschrift fuer die Mineraloelwirtschaft*, February 1975.)

does demonstrate the overwhelming effect of density on line loss and line capacity.

A supercritical line has other latent advantages. Its capacity can be increased by adding booster stations at a fraction of the energy needs of a subcritical line, and the line contains a much larger inventory which makes for a more stable operation.

Of great importance for the design of the line is the geodetic profile. In a supercritical line large differences in elevation will cause correspondingly large changes in the pressure gradient. Pressure minima will occur at peak elevations. Expansion from a higher to a lower pressure causes cooling, and compression causes heating of the gas; hence, with a line through mountainous territory, care must be taken that the limiting conditions of Table 8 are not trespassed at any point of the line. Figure 22, by way of example, shows the line pressure profile for the Wilton–Runcorn (ICI) ethylene line that crosses the Pennine hills at an altitude of 535 m across the narrow waist of England [51].

Ethylene lines are normally buried with 1.0 to 1.5 m of cover and must be

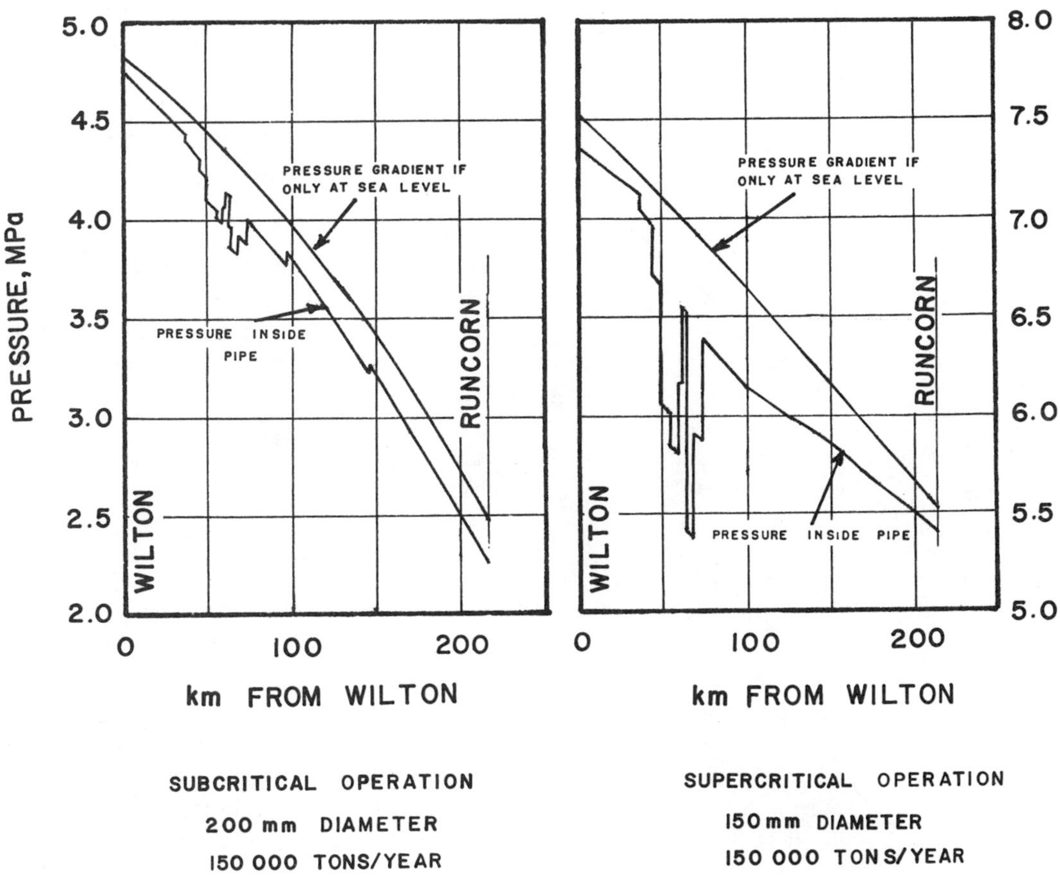

FIG. 22. Pressure gradients for Runcorn-Wilton ethylene pipeline.

protected with cathodic anticorrosion circuitry. Possible means of recouping the pressure energy at the receiving end of supercritical lines with one-directional flow may be installed whenever the volume of flow is adequate to make such a scheme profitable.

5.2.6.3. Commissioning of Ethylene Pipelines. Before a newly constructed ethylene pipeline can be placed in operation, it must be hydraulically tested for tightness, then drained and dried. With regard to safety as well as for carrying out the precommissioning procedures, the line is always subdivided by installing block valves at strategic and readily accessible points, and providing drain valves at low and vents at high spots. Filling and pressure testing is done by section over drain connections most conveniently located in respect to an adequate and acceptable supply of fresh water. During the hydraulic test the line is routinely inspected to insure tightness of welds and valves.

The drying operation is of utmost importance because of the very low permissible water content of the ethylene (10 ppm) and to avoid the formation of troublesome hydrates at high pipeline pressures.

After draining the line it is usual to wipe the line several times with tight-fitting swabbing pigs that are launched from pig locks installed in the line, using dry nitrogen as the pig propellant. As a last and final procedure, the line is wiped with a 10-m or so long plug of anhydrous methanol enclosed between two pig-scraper devices. If traces of methanol left in the line after this treatment exceed the permissible methanol limit (determined by chromatographic analyses of the purge gas between line segments), it is best to purge the line through all available valves successively with nitrogen at 2.5 to 3 MPa pressure.

5.2.6.4. Cost of Ethylene Pipeline Transport. The cost of pipeline transport of ethylene can readily be calculated only for one-directional flow between a producer and consumer plant, and exclusive of the cost of any storage facilities. In 1970, Jozon calculated supercritical transport costs as between 2.2 and 5.5¢/t/km for 600,000 and 100,000 t/yr of ethylene, respectively [51]. The investment in this instance was to be paid off at an 8% interest rate over a 10-yr period. Later, unpublished estimates by CE-Lummus showed that costs by 1976 had nearly doubled and stood at about 6.0¢/t/km for a volume equivalent to 250,000 t/yr of ethylene.

5.3. Handling of Ethylene

Complete information on the safety, toxicology, and handling of ethylene is presented in a Chemical Safety Data Sheet produced by the Manufacturing Chemists Association [55]. Ethylene is a colorless gas at room temperature and pressure. The gas has a mild odor and is not irritating to the eyes or respiratory system. Thus it does not give adequate warning of health or fire hazards. For purposes of controlling exposures to ethylene and preventing fire and explosion hazards, the concentration of ethylene should not be allowed to be greater than 5500 ppm.

When inhaled in sufficiently high concentrations, ethylene causes narcosis, and if oxygen is deficient, it will produce asphyxiation. Chronic injuries to humans resulting from exposure to ethylene have not been reported. In animals, testing to date has not demonstrated lasting or cumulative effects. However, detailed testing is being undertaken by the Chemical Industry Institute of Toxicology.

Ethylene is regulated by the U.S. Department of Transportation as a flammable compressed gas. Regulations and codes have been developed for the design of ethylene manufacturing, storage, and transportation facilities [55].

6. Manufacture

6.1. Raw Materials

Ethylene is almost exclusively produced from ethane, propane, and heavier paraffins and from the virgin crude oil fractions naphtha, kerosene, and gas oil. Occasionally, raffinates from aromatics extraction facilities are used as a supplementary raw material. Small quantities of ethylene are recovered from coke oven gas and from the gases produced by fluid catalytic crackers. Ethylene is also produced from crude oil directly [56–58], but this route to ethylene has, for a variety of technical and economic reasons, so far not gained commercial significance.

In the United States, ethane and propane have been the traditional feedstocks for the production of ethylene. These raw materials are recovered in natural gas processing plants of various capacities scattered throughout the gas-producing areas. Their main function is the conditioning of natural gas for pipeline transmission. The ethane so produced usually has a purity of approximately 95% with the remaining 5% being methane and propane in about equal amounts. Minor amounts of carbon dioxide and hydrogen sulfide are usually present. The LPG fractions (propane, butanes, and mixtures of both) produced by natural gas plants generally conform to NGPA [59] specifications. While nearly all recovered ethane is used for ethylene manufacture, substantial competing demands exist for propane and butanes as household and small industry fuel. Frequently blends of ethane and propane are used as raw materials for ethylene production.

In countries without significant natural gas production, i.e., up until recently all of Western Europe and Japan, naphtha has represented the most economic raw material. Typical full-range naphtha has an ASTM boiling range from 40 to 180°C and a density of approximately 0.7 g/cm^3. Occasionally, narrower cuts such as are produced in the prefractionation of naphtha for aromatics plants find use as feedstocks for ethylene plants.

Naphthas are generally characterized by their PONA (*p*araffins, *o*lefins, *n*aphthenes, and *a*romatics) analysis which, to an extent, indicates the suitability for ethylene production. While the conversion of individual hydrocarbons to ethylene and coproducts has been widely investigated, no complete analytical model exists that allows precise determination of the petrochemical potential of complex hydrocarbon mixtures. From experimental work it is known that iso-paraffins and alkyl-cyclopentanes yield less ethylene than *n*-paraffins and alkyl-cyclohexanes. Therefore the petrochemical potential of naphthas is often determined on the basis of previous industrial or pilot-plant experience with naphthas of the same crude oil origin.

Over the last few years kerosenes and gas oils have become economically attractive feedstocks, and several modern ethylene manufacturing facilities in the United States as well as in Europe are using naphthas and gas oils in various ratios as their total feedstock requirements. The yields of ethylene from gas oil are generally lower than the yields from naphtha, and the presence of polycyclic hydrocarbons reduces the value of gas oil as feedstock for the production of organic derivatives. The characterization of middle distillates by boiling range, density, and PONA analysis is identical with that of naphthas. Research on the pretreatment of these hydrocarbons and even vacuum distillates aims at an enhancement of the petrochemical yields (see Chapter 8).

In recent years much interest has also been shown in the production of ethylene directly from crude oil. BASF in Germany was the first to announce such a process (1958) employing partial oxidation by means of submerged combustion. Very recently UCC has disclosed [187] additional development work done in conjunction with Kureha Chemical and Chiyoda on this process. Process design work has been completed on a 2300 t/yr semicommercial plant to be erected at UCC's Seadrift, Texas, site by the third quarter of 1979. Active in this area is also the Dow Chemical Co. [184]. The objectives of these investigations are obviously the elimination of the crude distillation unit to prepare feedstocks and overcoming the coke problem. However, a partial oxidation process will require an oxygen plant. Concerned quarters in Europe, the United States, and Japan are awaiting with interest an economic comparison of this new technology with the current one employing pyrolysis heaters.

In 1970 approximately equal amounts of ethylene were produced worldwide from light paraffins (ethane and propane) as from naphtha [19]. In the United States, due to the accelerating shortage of natural gas, the relative importance of gas oil as the raw material for ethylene production is expected to increase significantly as shown in Table 9.

In Europe the first large-scale ethylene manufacturing facilities based on ethane and propane recovered from North Sea natural gas are presently being designed. Very large quantities of ethane and propane will become available in the oil-producing countries as a result of projects now under construction to stop the flaring of associated gas entirely.

Interest in ethylene production through dehydration of ethanol has been revived in the wake of the substantial price increases of the traditional hydrocarbon feedstocks. Ethanol can be produced from such renewable resources as grains (e.g., corn and wheat), potatoes, sweet sorghum, sugar beets and cane, by-products of sugar production (molasses), wastes from food processing industries, fruit juices, and various tropical fruits and roots by well-known fermentation/distillation processes, and potentially also from a wide range of cellulosic biomass materials (e.g., agricultural and municipal wastes, and forest sources) which first must be hydrolyzed to fermentable sugars by acids or enzymes. This route to ethylene offers the opportunity to establish a limited petrochemical industry on the basis of raw materials not derived from natural gas or crude oil, and as such has received much interest in countries that have but scant hydrocarbon resources [60]. Should this route become of industrial significance it could lead to the paradox that in some parts of the world ethanol would be dehydrated to ethylene while elsewhere ethylene would be hydrogenated to produce ethanol (see Chapter 3).

Coal can be processed by means of conventional and novel coal gasification and coal liquefaction techniques to produce feedstocks suitable for ethylene production. These routes to ethylene have so far not gained commercial significance with the exception of a plant in South Africa where ethylene is produced from coal on a limited scale. As coal becomes more important as a source of energy through new coal conversion technology, significant amounts

TABLE 9 Raw Materials for Ethylene in the United States (percent of ethylene produced from various raw materials)

	1970	1980
Ethane	49	42
Propane	30	24
Refinery gases	7	—
Liquids (naphtha and gas oil)	14	34

TABLE 10 Typical Pyrolysis Yields of Ethylene Plant Feedstocks (in mass percent): Light Paraffins (pure compounds)

	Ethane		Propane		*n*-Butane	Isobutane	*n*-Pentane	Isopentane
Conversion (%)	50	60	75	90	95	92	95	95
H_2	3.06	3.55	1.08	1.29	1.0	1.39	0.8	0.9
CH_4	2.6	4.2	18.83	24.7	21.8	23.6	17.52	20.6
C_2H_2	0.12	0.25	0.17	0.33	0.4	0.46	0.55	0.42
C_2H_4	41.65	48.2	25.0	34.5	35.8	10.3	36.5	17.95
C_2H_6	50.0	40.0	5.15	4.4	5.1	0.73	5.6	3.2
C_3H_4	0.1	0.02	0.22	0.34	0.55	2.46	0.95	1.7
C_3H_6	0.89	1.11	18.86	13.96	16.4	21.2	19.8	19.0
C_3H_8	0.14	0.17	25.0	10.0	0.15	0.33	0.55	0.55
C_4H_6	0.5	1.07	1.05	2.65	3.4	1.94	3.7	4.6
C_4H_8	0.25	0.21	1.28	1.0	1.7	16.7	2.6	14.38
C_4H_{10}	0.35	0.27	0.07	0.05	5.0	8.0	0.1	0.1
C_5s	0.2	0.27	1.43	1.81	1.65	2.12	6.1	7.5
C_6H_6	0.2	0.48	1.09	2.2	2.58	3.42	1.8	4.1
C_7H_8	0.03	0.06	0.33	0.48	1.12	1.51	0.6	1.25
C_8H_8					0.45	0.67	0.1	0.25
C_8H_{10}							0.06	0.15
C_6–C_8 PON		0.14	0.14	1.44	1.2	3.18	0.1	0.1
C_9–200°C							1.48	1.75
Fuel oil			0.3	0.85	1.7	1.99	1.09	1.5

of chemical products (including ethylene and ethylene derivatives) may eventually be derived from coal conversion operations. Any effect on the use of traditional feedstocks for ethylene, however, is not probable to be large in the near future.

6.2. By-products

Except in the case of ethanol dehydration which is so far of minor industrial significance, the manufacture of ethylene invariably entails the coproduction of industrially important petrochemicals. Typical yield distribution patterns for a variety of feedstocks are shown in Tables 10 and 11. Some of the by-products require further processing to obtain marketable petrochemical products.

6.2.1. Hydrogen

All ethylene manufacturing facilities produce significant quantities of hydrogen, yet only for large ethane-based ethylene plants is the coproduction of ammonia economically interesting.

TABLE 11 Typical Pyrolysis Yields of Ethylene Plant Feedstocks (in mass percent)[a] Petroleum Fractions

	Naphtha		Kerosene	Gas Oil	Vacuum Gas Oil
Severity	Medium	High	High	High	High
H_2	0.63	0.8	0.63	0.65	0.47
CH_4	11.7	15.3	11.4	10.6	7.9
C_2H_2	0.3	0.75	0.37	0.35	0.2
C_2H_4	24.0	29.8	23.2	24.0	19.5
C_2H_6	3.9	3.75	3.4	3.25	2.83
C_3H_4	0.76	1.1	0.85	1.0	0.38
C_3H_6	15.75	14.1	13.8	14.45	11.3
C_3H_8	0.41	0.25	0.45	0.45	0.36
C_4H_6	4.6	4.85	4.4	4.7	5.5
C_4H_8	6.85	4.2	4.0	4.5	4.88
C_4H_{10}	1.0	0.25	0.1	0.1	0.09
C_5s	5.0	2.3	3.0	3.3	5.0
C_6H_6	3.15	6.3	7.2	5.8	4.16
C_7H_8	4.4	4.9	4.0	3.15	2.5
C_8H_8	2.65	2.0	2.2	1.2	1.28
C_8H_{10}	0.52	0.7	0.75	0.7	0.5
C_6–C_8 PON	8.83	2.0	1.9	1.8	2.0
C_9–200°C EP	3.75	2.8	4.85	2.4	2.0
Fuel oil	1.8	3.8	13.5	17.6	29.15

[a]Valid only for feedstocks listed in Table 12.

The hydrogen product of ethylene plants usually has a purity of 95 mol% with methane as the major contaminant. This product generally contains less than 5 ppm of carbon monoxide and is free of sulfur. A relatively small portion of the hydrogen product is used within the ethylene plant for various hydrogenation steps while the remainder is available as chemical hydrogen for utilization elsewhere in refineries and/or within petrochemical complexes.

6.2.2. *Acetylene*

Acetylene can be a minor by-product of the typical ethylene production facility. Most often acetylene is hydrogenated to ethane within the ethylene plant which is then recycled to the heaters to yield additional ethylene. Only liquid feedstocks yield economically significant quantities of acetylene that are typically in the order of 1 to 2% of the ethylene production. The recovery of high-purity by-product acetylene can be very attractive in instances where a demand for only moderate amounts of acetylene exists [61]. Some large modern ethylene plants are designed for the recovery of high-purity acetylene.

6.2.3. *Methylacetylene and Propadiene*

These compounds are of little interest to the chemical industry. Their yield from liquid feedstocks is in the range of 1 to 2% of the ethylene yield. Dow Chemical Co. has pioneered the production and marketing of these unsaturated hydrocarbons as welding gases. They are sold as a substitute for acetylene, offering the advantages of lower cost and greater ease and safety of handling.

6.2.4. *Propylene*

Propylene is the major by-product of all ethylene plants using feedstocks other than ethane. In fact, the production of propylene from ethylene plants is so important that the name olefins plants for these manufacturing facilities seems more appropriate and is often preferred. Propylene production typically amounts to between 40 and 70% of the ethylene production. It is recovered as either so-called chemical grade (90 to 95%) or polymerization grade (99.6%). Either product must be essentially free of diolefins, acetylenes, and other olefins (see Table 13).

6.2.5. *Butadiene and Butenes*

Ethylene plants based on naphtha and heavier feedstocks produce significant quantities of mixed C_4-hydrocarbons, predominantly 1,3-butadiene and isobutylene. This by-product represents an important raw material for the rubber industry which converts it into styrene–butadiene, polybutadiene, and polyisobutylene elastomers, acrylonitrile–butadiene–styrene copolymers, and similar materials.

With the expected shift of ethylene plant feedstocks in the United States from natural gas liquids to naphtha and gas oil, the butadiene supply is expected to increase significantly [62] with an attendant decrease of butadiene imports from Europe and Japan and the decline of butadiene production by butane dehydrogenation. The projected abundant supply of butadiene and butenes may lead to new conversion routes to other petrochemicals [63].

6.2.6. *Pyrogasoline*

Significant quantities of pyrogasoline are produced by ethylene plants which are based on naphtha and gas oil feedstocks. Pyrogasoline is a complex mixture of hydrocarbons ranging from C_5-compounds through material with a boiling point of approximately 200°C. It is rich in single ring aromatics, mono- and diolefins, and also contains a small but significant amount of isoprene.

Raw pyrogasoline produced by ethylene plants is highly unstable due to its high content of diolefins. Hydrogenation units are therefore closely associated with ethylene plants which produce significant quantities of pyrogasoline. In these units depentanized or full-range pyrogasoline is treated with hydrogen under partial pressures of 2.5 to 4.5 MPa over nickel or noble metal catalysts for the conversion of diolefins to monoolefins. Hydrotreated pyrogasoline has a Research Octane Number in the range of 90 to 100 and is a valuable motor gasoline blending stock.

For the recovery of aromatics, pyrogasoline after hydrogenation of diolefins is fractionated to obtain an aromatics-rich heart cut which is then subjected to a second more severe hydrotreating step for the complete saturation of all olefins. The product is then further processed in aromatics extraction or hydrocracking units. Aromatics extraction produces benzene, toluene, xylenes, and ethylbenzene while hydrocracking of the aliphatics and hydrodealkylation of alkylbenzenes produces high-purity benzene. In one such hydrocracking process the above-mentioned second hydrotreating step is combined with the hydrocracking operation.

The hydrogen product of ethylene plants is in most but not all cases of sufficient volume to satisfy also the hydrogen requirements of the aromatics recovery facilities.

6.2.7. *Pyrolysis Fuel Oil*

Negligible amounts of fuel oil are produced by plants producing ethylene from light paraffins. Plants using naphtha as raw material produce only modest amounts (10 to 20% of the ethylene production) of fuel oil which is conveniently burned in fired equipment within the plant. The amount of fuel oil produced by

TABLE 12 Feedstocks for Ethylene Plants: Typical Characteristics of Petroleum Fractions

	Naphtha	Kerosene	Gas Oil[a]	Vacuum Gas Oil[a]
Origin	Kuwait	North Sea	Kuwait, North Sea	Arabian Light
Density, g/cm^3	0.724	0.801	0.839	0.894
Sulfur content, mass%	0.04	0.03	0.1	0.2
ASTM boiling range (°C):				
IBP	28	180	220	274
50%	121	206	290	453
EP	175	242	380	555
BMCI	—	29.7	27.7	35.4
PONA analysis (mass%):				
Paraffins	72.3	50.9	41.8	17.5
Olefins	—	—	—	2.0
Monocyclic naphthenes	16.7	15.8	19.2	11.2
Polycyclic naphthenes	—	10.5	14.1	21.4
Monocyclic aromatics	11.0	14.9	20.9	25.0
Polycyclic aromatics	—	7.9	4.0	22.9

[a]After hydrodesulfurization.

plants using gas oils is substantial, and with vacuum gas oil feedstocks it will exceed the ethylene production (see Table 11).

Pyrolysis fuel oil is a highly unstable material and generally unsuited for disposal in admixture with other fuel oils. Burning it in boilers or large process heaters with a minimum of intermediate storage represents the preferred method of disposal.

The feedstocks of ethylene plants are either naturally low in sulfur or are hydrodesulfurized. As a result, the pyrolysis fuel oil product of ethylene plants has a sulfur content of generally less than 1 wt.%. Low sulfur pyrolysis fuel oils having a high content of condensed aromatics are suitable raw materials for the production of carbon black and needle coke. Special technology has been developed to permit the delayed coking of pyrolysis fuel oil despite its high reactivity [64]. Depending on the specific processing route chosen, the resultant coke is either of the amorphous or the high quality needle type. The former is suitable for the production of anodes for the aluminum industry while the latter is converted to electrodes for the steel industry. In 1978 this technology has been applied on an industrial scale. Shell Nederland Chemie B.V.'s approach to by-product utilization at its Moerdijk, Netherlands, facility was to integrate its 500,000 t/yr ethylene unit with a 100,000 t/yr needle coke plant [217].

6.3. Typical Feedstock Characteristics

Table 12 shows typical characteristics of petroleum fractions. These should not be considered rigid specifications for feedstocks but rather as representative of feedstocks to ethylene plants presently in operation. So far little vacuum gas oil is used for the production of ethylene, and where it is used the end point is generally not as high as shown. Occasionally, raffinates of reformates or wide boiling-range natural gas condensates with characteristics quite different from those listed in the table are used as raw materials.

In addition to the feedstock characteristics listed to appraise the suitability of a feedstock for ethylene manufacture, the Bureau of Mines Correlation Index (BMCI) is used to estimate yields of ethylene from liquid stocks. While it is not infallible, in certain instances of single components having the same BMCI value yet differing widely in yield potential it has proven surprisingly useful in correlating ethylene yield data for liquid feedstocks of different origin.

The *B*ureau of *M*ines *C*orrelation *I*ndex is defined by the following formula:

$$\text{BMCI} = \frac{48640}{\text{VABP}} + 473.7(d_{15}) - 456.8$$

VABP is the *v*olume *a*verage *b*oiling *p*oint (°K), and (d_{15}) the density in g/cm^3 at 15°C.

The BMCI is a measure of the extent of the naphthenic and aromatic ring structure of a hydrocarbon feedstock and is calibrated to yield values of 0 and 100 for pure *n*-hexane and pure benzene, respectively.

TABLE 13 Typical Specifications for Products of Ethylene Plants

Ethylene product	
Ethylene	99.9 mol% min
Methane and ethane	0.1 mol% max
Hydrogen	10 mol ppm max
Acetylene	2 mol ppm max
Propylene and propane	10 mol ppm max
Oxygen	1 mol ppm max
Carbon monoxide	2 mol ppm max
Carbon dioxide	5 mol ppm max
Water	10 mol ppm max
Methanol	5 mol ppm max
Total sulfur	1 mass ppm max
Propylene product	
Propylene	99.6 mol% min
Paraffins	0.4 mol% max
Hydrogen	10 mol ppm max
Ethylene	5 mol ppm max
Acetylene	1 mol ppm max
Methylacetylene	5 mol ppm max
Propadiene	5 mol ppm max
C_4-Hydrocarbons	5 mol ppm max
Carbon monoxide	5 mol ppm max
Carbon dioxide	5 mol ppm max
Methanol	5 mol ppm max
Oxygen	2 mol ppm max
Water	10 mass ppm max
Sulfur	2 mass ppm max
Hydrogen product	
Hydrogen	95 mol% min
Methane, ethylene, and ethane	Balance
Hydrogen sulfide	1 mass ppm max
Carbon monoxide	5 mol ppm max
Oxygen	5 mol ppm max
Water	5 mol ppm max
Acetylene product	
Acetylene	99.5 mol% min
Ethylene	0.4 mol% max
Methane	0.05 mol% max
Methylacetylene and propadiene	0.01 mol% max
Total C_3-hydrocarbons	0.1 mol% max
Hydrogen	20 mol ppm max
Oxygen	10 mol ppm max
Carbon monoxide	10 mol ppm max
Carbon dioxide	10 mol ppm max
Solvent	5 mol ppm max
C_4-hydrocarbons	
C_3-hydrocarbons	0.1 mol% max
C_5-hydrocarbons	0.1 mol% max
Total sulfur	5 mass ppm max

(*continued*)

TABLE 13 (*continued*)

Composition[a]		
C_3-hydrocarbons	0.1	mol%
1,3-Butadiene	35–50	mol%
Ethylacetylene	0.2	mol%
Vinylacetylene	0.8	mol%
cis-Butene-2	4–5	mol%
trans-Butene-2	4.5–6.5	mol%
Isobutene	20–30	mol%
1-Butene	15–25	mol%
n-Butane	5–15	mol%
Isobutane	0.5–1.0	mol%
C_5-hydrocarbons	0.1	mol%
Pyrogasoline		
C_4-hydrocarbons	0.5	mass% max
ASTM distillation end point	200°C	
Composition[a]		
C_4-hydrocarbons	0.5	mass%
C_5-hydrocarbons	10–20	mass%
Benzene	20–35	mass%
Toluene	15–25	mass%
C_8-aromatics	10–15	mass%
C_6–C_8 PON	5–35	mass%
C_9–EP	10–20	mass%

[a]The composition is dependent on the composition of the pyrolysis feedstock and the pyrolysis operation.

6.4. Yields

Tables 10 and 11 show typical single pass pyrolysis yields for pure light paraffins and liquid petroleum fractions, respectively. Ethane which remains unconverted or is formed through the pyrolysis of other feedstocks can be, and usually is, further pyrolyzed to extinction, thereby increasing the ultimate ethylene yield from any given feedstock. For some important feedstocks, two yield spectra are shown which represent economically interesting alternate operating modes for pyrolysis reactors.

Again the yields shown should be considered typical; to what extent they can be influenced by the design and operation of the pyrolysis reactors or through pretreatment of feedstocks is discussed in Chapter 8.

6.5. Product Specifications and Characteristics

Table 13 shows typical specifications and characteristics for all products of ethylene plants. The ethylene product invariably has to be of high purity with stringent limits on the concentration of certain impurities. The same applies to the propylene product if polymerization quality is required. For the last 15 years the licensors of the polymerization processes have imposed increasingly tighter specifications for the two main products of ethylene plants.

The other products have to meet mostly fractionation specifications although an accurate prediction of their composition is generally required for the design and operation of subsequent processing facilities. The pyrolysis fuel oil usually has to meet only a flash point specification of 70°C minimum.

6.6. Manufacturing Processes

6.6.1. Pyrolysis of Hydrocarbons

The large-scale production of ethylene is carried out by pyrolysis (cracking) of hydrocarbon feedstocks in the presence of steam in large, multiple pyrolysis furnaces and subsequent separation of the resultant gas mixture through a complex sequence of unit operations.

The original ethylene manufacturing process, as described by Pratt and Foskett in an early paper [65], has not changed fundamentally over the last 30 years. Yet a number of evolutionary technical developments coupled with a more than tenfold increase in single plant capacities have brought about a continuous decrease of the total resource requirements to produce a unit amount of ethylene. Per unit of production capacity, the modern ethylene plant contains less steel and other materials, consumes less feedstock and energy, and requires less operating manpower, given a comparable feedstock, than did the typical plant of 15 years ago. If it were not for the recent, unusually steep increase in the cost of hydrocarbons and capital goods, the modern ethylene plant would also produce ethylene at a lower cost per unit of production than did older plants.

Technological advances resulted in the vertical tube pyrolysis furnace, with coils made of novel alloys which offer longer life at higher operating temperatures. New furnace insulating materials and methods were developed to accommodate higher fire box temperatures. The ethylene production capacity of a single furnace has increased from 10,000 t/yr in 1956 to over 50,000 t/yr in 1976, and the thermal efficiency of a modern pyrolysis furnace is exceeding 90%. The modern pyrolysis furnace also has the function of preheating boiler feedwater and superheating high-pressure steam. Including its appurtenant transferline exchangers, the entire pyrolysis module, in fact, represents a combination of pyrolysis reactor and steam generator (see Section 8.6).

The uncertainty surrounding feedstock availability and cost has caused the development of pyrolysis heater designs suitable for the processing of both naphtha and gas oil. In 1978, several modern ethylene plants were in operation with such feedstock flexibility.

Increasing plant capacities led to the introduction of steam turbine driven centrifugal compressors. The excellent performance record of this machinery in turn has led to the widespread if not unanimous acceptance of the single line compression concept in place of multiple reciprocating compressors.

The cryogenic treatment of the light pyrolysis gases became simpler and more efficient when multipass brazed aluminum exchangers of sufficient capacity and adequate design pressures became available. High heat flux tubing for heat exchangers in condensing-vaporizing service, a relatively recent development of the Union Carbide Corp. [66, 67], offers savings in heat transfer area and/or energy requirements for the large condensers and reboilers in ethylene plants.

With increasing plant capacities, equipment became so large that their positioning on various levels in structures was either no longer practical or economical. This led to plant layouts with most equipment at grade level but it also called for the application of cryogenic reflux pumps.

Much development work has been aimed at the decrease of the energy requirements of ethylene plants, mostly through more extensive recovery of heat and cold which, to some extent, can only be accomplished at the expense of increased integration and complexity of the process.

In the last few years the digital process control computer has found increasing acceptance by the petrochemical industry. Many modern ethylene plants contain process computers with tasks ranging from simple data and alarm logging to full supervisory control for optimum performance of the entire plant. The present rate of development in the mini- and microcomputer field is likely to bring further significant advances of the scope and capability of process control systems. In ethylene plants, as in other plants of comparable complexity, a higher level of automation will tend to increase profitability in the future.

Capacities of single train ethylene plants seem to be still on the increase. A 750,000-t/yr plant is presently under construction in the United States. While such large single train plants are technically feasible at the present state of the art, their operation involves increased economic risks from exposure to fluctuating market conditions. Projects of this size may be considered by the few very large ethylene producers with a high captive demand for ethylene, particularly under circumstances when some existing plant capacity is about to be retired because of obsolescence. However, ultralarge plants (750,000 t/yr and over) will be the exception, while large plants (in the 400,000 to 600,000 t/yr range) are more likely to be built in greater numbers.

6.6.2. Ethanol Dehydration

The chemistry of ethanol dehydration dates back to the eighteenth century when it was first reported that ethylene was released by passing ethanol or ether over heated alumina or silica [68]. Since then much research has been devoted to the investigation of various catalysts [69] although only activated alumina and phosphoric acid on a suitable support have become of industrial importance.

To this day the actual reaction mechanism of ethanol dehydration is not fully understood. It has been postulated that the dehydration proceeds partially directly to ethylene but also involves the formation of ether as an intermediate reaction product.

$$C_2H_5OH \longrightarrow C_2H_4 + H_2O \tag{22}$$

$$2C_2H_5OH \longrightarrow (C_2H_5)_2O + H_2O \longrightarrow 2C_2H_4 + 2H_2O \tag{23}$$

The formation of ether is favored at temperatures of approximately 230°C while at temperatures of 300 to 400°C ethylene with a minimum content of ether is obtained. The practical ethylene yield is approximately 94% of the theoretical value. Further purification is required to remove traces of aldehydes, acids, higher hydrocarbons, and carbon dioxide as well as the water formed.

Ethanol dehydration was widely practiced in Europe on a small industrial scale before low-cost ethylene became available from hydrocarbon-based ethylene plants. In 1978, small plants were still in operation in Brazil, India, Pakistan, and Peru. Brazil and India have developed ambitious plans to increase the production of ethylene from ethanol as a means of using domestic resources rather than imported hydrocarbons as raw materials for their petrochemical industries.

Figure 23 presents a simplified processing scheme of a commercial ethanol-based ethylene fixed-bed process.

The ethanol feed is pumped into a steam-heated vaporizer. The ethanol vapors are preheated with high-pressure steam and passed over a catalyst of specially treated activated alumina for dehydration. A high yield of ethylene, approximately 94% of theoretical, is produced in one pass, 1% remains unconverted, and the remainder is converted into undesirable by-products.

Sufficient reaction volume is provided to achieve essentially complete conversion of the ethanol feed and the ether formed. This reduces the yield of ether to a minimum and avoids the necessity of recovering and recycling ethanol.

The heat required for the reaction is supplied by condensing Dowtherm vapor in the reactor shell. Temperature control is important. If the temperature is too high, aldehydes are formed; if it drops too low, ethers are obtained. The reactor is designed to distribute heat evenly and to avoid hot spots. The catalyst is regenerated with air and steam every few weeks to remove carbon deposits.

The ethylene-rich gaseous reactor effluent is quenched and water washed to cool the vapors and at the same time remove the unreacted alcohol and traces of aldehydes and acids. This is followed by washing with dilute caustic soda to remove carbon dioxide. The gas is then sent to a gas holder.

The gas from the gas holder is compressed, cooled by a refrigeration system, and sent first through an adsorber containing a bed of activated carbon, which selectively adsorbs "heavy" impurities such as butane and butylenes, and then through a second vessel containing a desiccant, which removes water vapor from the ethylene. Dual systems are provided to permit continuous operation during reactivation of the adsorbants.

The purified ethylene is dry and has a purity in excess of 99 mol%. It meets the customary specification for polymerization-grade ethylene.

Ethanol containing 5% water is suitable as feedstock. If denaturated alcohol is used, 5 wt.% of ethyl ether as denaturant is preferred.

Fluid-bed processes, in place of the fixed-bed externally heated reactor systems, have been investigated extensively but have so far not found industrial application.

Fluidized-bed systems have a number of highly useful characteristics, namely ease of temperature control, improved heat and mass transfer, continuity of operation, and the ability to process large tonnages of reactants.

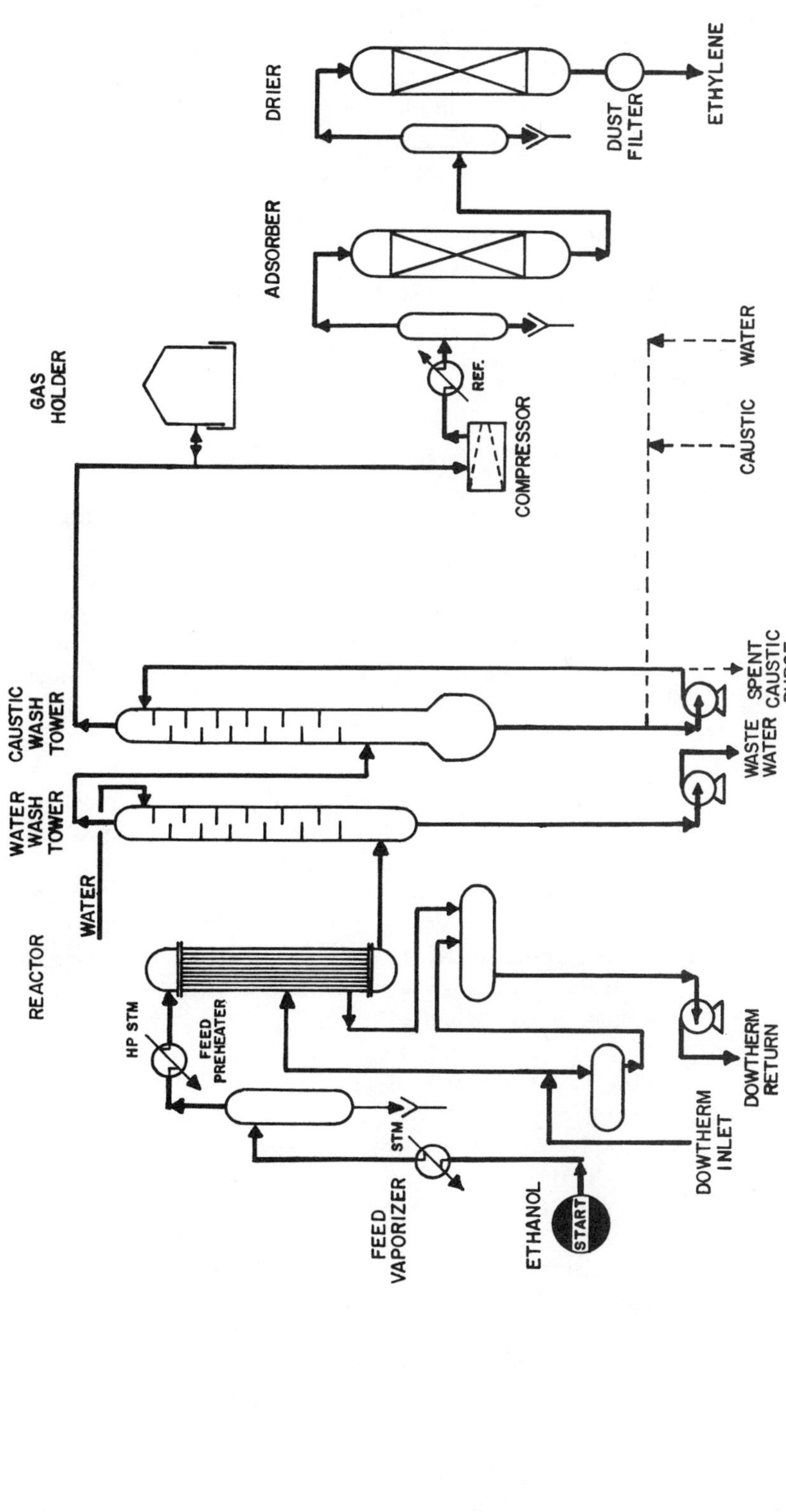

FIG. 23. Ethylene from ethanol process.

Instead of multiple fixed bed reactors (five, in the case of a 60,000 t/yr ethylene plant, for example) together with their auxiliary equipment, only a single fluidized-bed reactor and regeneration system are required. The required heat for the reaction is supplied by circulating the fluidized catalyst through the regenerator. Just as in the fixed-bed process, close temperature control is important for the ethanol dehydration reaction and is easily achieved in a fluidized bed. In a fluidized system a uniform temperature is quickly established throughout the fluidized bed without formation of hot or cold spots.

Ethylene yields of 99% are obtained as compared to 94 to 96% with fixed-bed systems. Besides close temperature control, high yields are also obtained by keeping the catalyst at high activity at all times. This is accomplished by circulating the catalyst through a regenerator and by replacing catalyst which is lost due to attrition. Preliminary estimates indicate that an ethanol dehydration plant employing a fluidized-bed process will require a substantially lower capital investment than does a plant using fixed-bed reactors [211].

6.6.3. *Ethylene from Coal*

There are several possible routes from coal to ethylene by conventional and by novel technology.

Synthesis gas produced from coal by commercially proven coal gasification processes (Lurgi, Koppers-Totzek, Winkler) may be converted into hydrocarbons by Fischer-Tropsch (FT) processes. Historically, FT reactions have produced higher molecular weight saturated and olefinic hydrocarbons as well as oxygenated compounds. Ethylene may be produced by cracking of the hydrocarbons (which may also be produced by second generation coal liquefaction processes) and by dehydration of recovered ethanol. Some ethylene may also be recovered directly. Conventional FT technology, however, is presently not economically viable for ethylene production.

Today, the only FT plant operating is in South Africa at Sasolburg. The plant, now referred to as Sasol I, was built in 1955 and employs Lurgi gasifiers to produce synthesis gas from low-grade coal which is converted into hydrocarbons by two different FT processes, a fixed-bed process (ARGE) and a fluid-bed process (Synthol). Sasol claims that the combination of the two systems yields virtually the full range of products (including ethylene) normally obtained from petroleum. A similar new but much larger facility, Sasol II, is now planned to be on stream by 1981 [212]. In contrast to other industrialized countries, in South Africa FT production is economically viable for two reasons: (1) the large coal deposits can be mined at extremely low costs, and (2) South Africa is striving to become independent of oil imports [213].

After the 1973 oil crisis, coal was reexamined as an important source for fossil carbon by many industrialized countries. Considerable activity is currently in progress to develop coal-based processes for the production of ethylene. The goal is to free the petrochemical industry from total reliance on crude oil and natural gas. The probable advent of large-scale first and second generation coal gasification and coal liquefaction process units in the United States and possibly other countries by the late 1980s for fuel purposes has reawakened interest in coal-derived synthesis gas and hydrocarbons as petrochemical feedstock.

Depending on reaction conditions (catalyst, temperature, pressure), synthesis gas can be converted to many different products. Efforts directed toward optimization of this technology for ethylene production were summarized by Brownstein [212]. More recently Ruhrchemie has developed a new FT iron catalyst modified by activators that can convert synthesis gas directly to lower olefins, primarily ethylene and propylene, in good yield [213]. The process produces propylene in about the same ratio to ethylene as in a conventional steam cracking plant.

In addition to direct ethylene production, indirect processes for ethylene manufacture have also been proposed which are based on coal-derived methanol.

In one process ethanol is produced by methanol homologation with synthesis gas followed by ethanol dehydration [212].

Mobil has developed a process for the conversion of methanol primarily to ethylene and propylene using a zeolite catalyst promoted with trimethylphosphite. At 620°C and a pressure of 2.5 to 5.0 MPa, ethylene and propylene yields of 18.9 and 36.3 wt.% are obtained, respectively, at 98% methanol conversion [214].

Today the routes from coal via coal conversion processes to reactive olefins for petrochemicals are still too expensive compared with their production from natural gas liquids or crude oil fractions. Nevertheless, as the supply of the latter is not inexhaustible, coal will become more important both as a fuel and raw material for the synthesis of petrochemicals. It is safe to say that the transition from one to the other will be slow and gradual, and will be largely dictated by technological improvements that will reduce capital and operating expenses of coal conversion processes.

6.6.4. Propylene Disproportionation

Phillips' triolefin process [215] converts propylene to ethylene and *n*-butenes by disproportionation. It is a means for altering the product distribution of refineries, ethylene plants, and other chemical plants. For example, if propylene is produced in a plant in excess of the quantity that can be used or sold, the excess amount can be disproportionated to produce ethylene and butenes, which may be more readily usable or salable. A commercial plant using the process to produce ethylene and butenes from 33,400 t/yr of propylene was built by Gulf Oil Canada Ltd. in 1966.

7. The Modern Ethylene Plant

The large-scale manufacture of ethylene from hydrocarbon raw materials is carried out in plants which are among the most complex and expensive facilities of the hydrocarbon processing industry. An ethylene manufacturing facility often forms the core of an entire petrochemical complex designed to produce a variety of petrochemical products.

The trend toward larger ethylene plant capacities is most pronounced in the United States where in 1978 several plants with capacities in excess of 500,000 t/yr are in operation while still larger plants are on the drawing boards. In Europe, Japan, and Brazil the petrochemical industry appears to have opted for ethylene plant capacities in the range of 300,000 to 500,000 t/yr with little discernable interest in larger capacities. Smaller plants are in operation and under construction in developing countries to serve smaller local markets.

Many modern ethylene plants are designed to process naphtha and gas oil in various ratios; some are designed for full naphtha–gas oil flexibility. A few plants can handle light vacuum gas oils with distillation end points as high as 450°C.

Plants to produce ethylene from light paraffins, particularly ethane and propane, are expected to be built in areas with abundant natural gas supply. In 1978 it remained an open question to what extent the growing trade of liquefied natural gas would spur the development of ethylene production in conjunction with either the liquefaction or vaporization of natural gas. The utilization of the refrigeration potential in liquefied natural gas for the cryogenic separation of ethylene from pyrolysis gases has been proposed for LNG receiving terminals [70].

All ethylene manufacturing processes incorporate a great number of unit operations; catalytic and noncatalytic reactions, absorption and adsorption, fractionation, compression, heat exchange, and phase separation are arranged in a complex array to achieve the production of high-purity ethylene and by-products in the most efficient manner.

Ethylene plants employ extreme temperatures which range from 850°C at the outlet of the pyrolysis furnaces to −165°C in the cryogenic hydrogen–methane separation system. These extreme temperatures call for the application of a multitude of materials of construction which include, besides the customary carbon steel, the high chrome-nickel alloys of the pyrolysis tubes; the stainless steel of the preheat sections in the pyrolysis furnaces; stainless steel, various nickel alloys, and aluminum in the cryogenic systems; and molybdenum alloy for the methanator which is exposed to high hydrogen partial pressures at elevated temperature.

Ethylene plants do not employ extreme pressures; except in the steam system, pressures generally do not exceed 4 MPa and nowhere in the process system are they below atmospheric pressure. Under steady-state conditions air can therefore not enter the process equipment nor are there any reaction systems, other than the fired heaters, which call for the continuous controlled reaction of hydrocarbons with oxygen. Certain equipment, notably the pyrolysis furnaces and hydrogenation reactors, have to be regenerated periodically to remove carbon or polymer deposits through controlled combustion. These operational steps are potentially hazardous. Ethylene plants contain a significant inventory of hydrocarbons and particularly light liquefied gases under elevated pressures, the inventory typically being equivalent to an accumulation of several hours. The safe containment of this inventory under normal and abnormal process conditions is of major concern to the designer and operator alike [71].

Ethylene plants usually contain one or more exothermic reaction systems in which acetylenic and diolefinic compounds are selectively hydrogenated to olefins or paraffins. Loss of control of the hydrogenation reaction has led to serious accidents in a few plants [72].

The typical ethylene plant contains well over 400 pieces of major equipment, i.e., fired heaters and heat exchangers, reactors, drums, fractionating towers, compressors, and pumps. For certain services a single piece of equipment cannot be economically manufactured with sufficient capacity and therefore multiple specimens are employed. This applies in particular to the pyrolysis furnaces, of which the typical ethylene plant contains eight to twelve, and to certain heat exchangers. Because of their maintenance requirements, pumps and reciprocating compressors are usually fully spared as is equipment such as the hydrogenation reactors which require periodic regeneration. Generally not spared are the large steam turbine driven centrifugal compressors for the pyrolysis gas and the two fluid refrigeration cascades. These have a combined energy consumption of approximately 50 MW for a 500,000 t/yr ethylene plant, which is unusually high for a hydrocarbon processing plant.

The complexity of the ethylene process and the large number of pieces of equipment require an extensive and equally complex automatic control system. The typical plant contains well over a thousand measuring instruments for temperatures, pressures, levels, flow rates, and compositions. It also contains hundreds of control loops. Most modern plants use electronic rather than the earlier pneumatic instrumentation, and the dedicated process control computer will play an increasingly important role in the efficient operation of future plants.

The efficient and reliable operation of ethylene plants is dependent on the performance of a great number of support systems which generally require an investment in the order of 40 to 70% of the cost of the process facilities. Utilities, storage and shipping facilities, pollution abatement, flare and safety systems, maintenance shops, and warehousing of spare parts are all needed to support the production of ethylene. In a plant feeding on naphtha or gas oil the steam system with its highest pressure level of 12 MPa and a temperature of 500 to 520°C is unusual for a hydrocarbon processing plant. Depending on the type of feedstock used, large quantities of steam, which may exceed the total steam requirements of the plant, are generated from the pyrolysis gases at the outlet of the pyrolysis furnaces and from flue gases over the convection section. Supplemented if necessary with steam from a conventional boiler, this steam is used to drive the major compressors and to satisfy all heat requirements of the process and support systems. Between 5 and 7 t of steam are consumed to produce 1 t of ethylene.

The extreme steam conditions call for fully demineralized boiler feed water which is produced in units similar to those found in large power stations. These units also contain facilities to polish the steam condensate. In fact, the boiler feed water and steam specifications [73] of the utilities industry are usually applied to the steam system of ethylene plants.

The optimization of the entire energy system of ethylene plants continues to be of much interest to the industry [74].

7.1. Ethylene Process Technology

Ethylene manufacturing technology represents an essentially open art and, unlike some processes such as ammonia synthesis, no company or individual inventors are singularly credited with any specific breakthrough in the art of ethylene manufacture. Most ethylene plants are based on technology and know-how developed by a number of engineering companies in the United States and Europe. A few engineering companies possess only technology for the design and engineering of pyrolysis furnaces, and others only for the separation and purification of the pyrolysis furnace effluents. Some companies own technology for complete ethylene manufacturing plants. A few petrochemical producers in the United States and Japan have elected to develop their own technology which they practice in their own manufacturing facilities.

Most of the available ethylene processes are basically similar to one another, which is an indication of the maturity that the technology has reached. Figure 24 shows the 19 major process systems of an ethylene plant designed to produce high-purity ethylene and propylene from naphtha or gas oil. Not shown is the refrigeration cascade, typically consisting of the thermally coupled propylene and ethylene refrigeration systems. This block flow diagram is self-explanatory; the individual systems are further discussed in subsequent sections.

Plants processing ethane feedstock are considerably simpler, usually containing neither secondary nor tertiary waste heat recovery systems nor producing fuel oil or pyrogasoline. The condensate deethanizer and the propylene and C_4 processing facilities are not required since the production rates of C_3 and heavier hydrocarbons are low.

Most modern plants provide for the superheating of the high-pressure steam in the convection section of the pyrolysis furnaces to reduce the total quantity of steam. Many plants, however, contain one or two separately fired steam heat superheaters, the absorbed heat duty of which adds directly to the steam production. The tertiary heat recovery system which produces quench water with a temperature of 80°C is only a true heat recovery system if the hot quench water is applied to the superfractionation of propylene and propane for the production of high-purity propylene. Where the recovered heat cannot be so used, this system is but a means of rejecting low level heat to the environment.

Most differences among various processes exist in the sequence of unit operations. The acetylene hydrogenation step is sometimes carried out at one of the compressor interstage levels or after compression. For plants processing ethane and/or propane, this results in the loss of most of the butadiene contained in the pyrolysis gas through hydrogenation along with the lighter acetylenic and diolefinic compounds. In this instance the loss is usually not important. For plants processing naphtha and heavier feedstocks, this front end hydrogenation step requires early separation of the C_4 and heavier material from the pyrolysis gas to preserve the butadiene.

In plants that are to produce high-purity acetylene as a by-product, acetylene recovery facilities take the place of the acetylene hydrogenation system, as shown in the block flow diagram, or treat the overhead from the ethylene–ethane fractionator. The former location offers the advantage of lower concentrations of acetylene in the high-pressure systems but requires handling of a larger gas volume and allowance for the presence of small amounts of C_3-compounds.

Little disagreement exists among designers about the need for early rejection of fuel oil and heavy gasoline in the processing of pyrolysis gases.

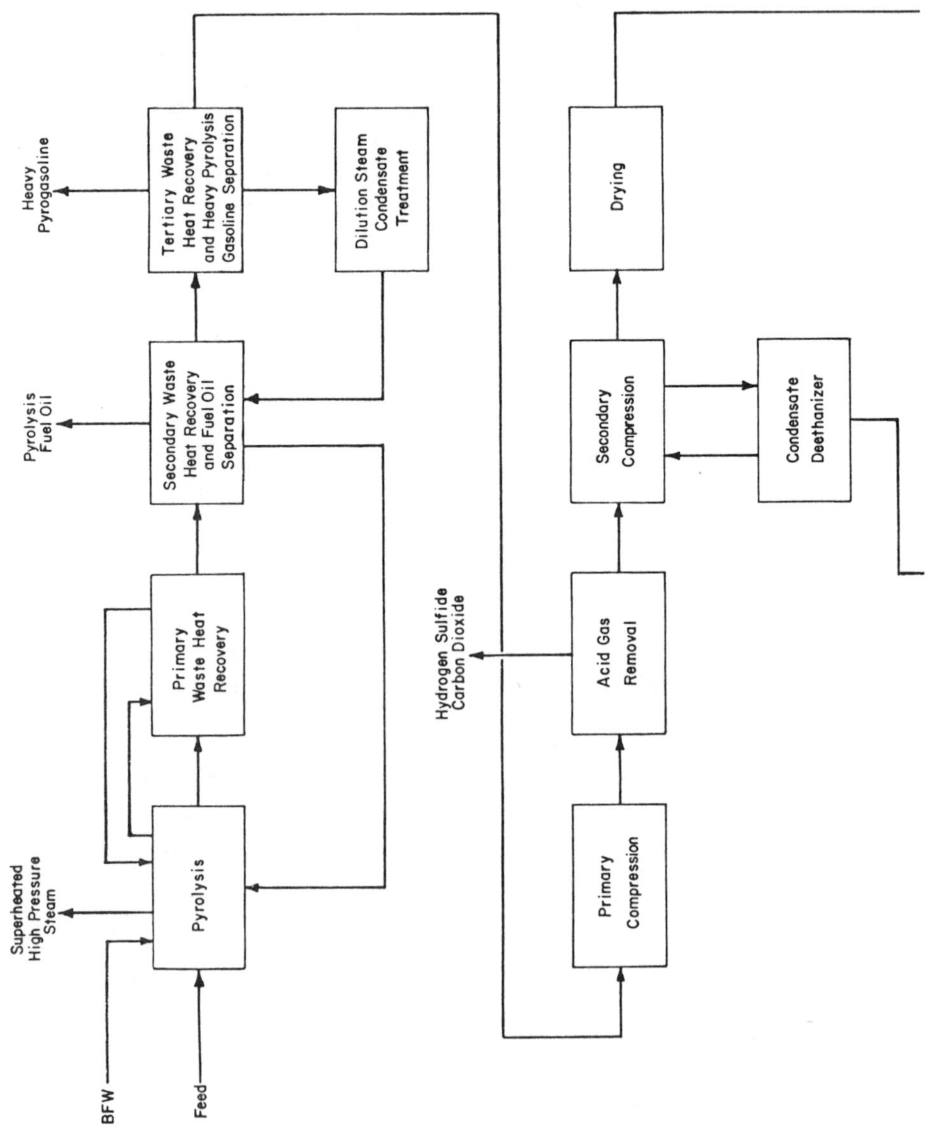
Heavy Pyrogasoline
Pyrolysis Fuel Oil
Superheated High Pressure Steam
Tertiary Waste Heat Recovery and Heavy Pyrolysis Gasoline Separation
Dilution Steam Condensate Treatment
Secondary Waste Heat Recovery and Fuel Oil Separation
Primary Waste Heat Recovery
Pyrolysis
BFW
Feed
Hydrogen Sulfide Carbon Dioxide
Drying
Secondary Compression
Condensate Deethanizer
Acid Gas Removal
Primary Compression

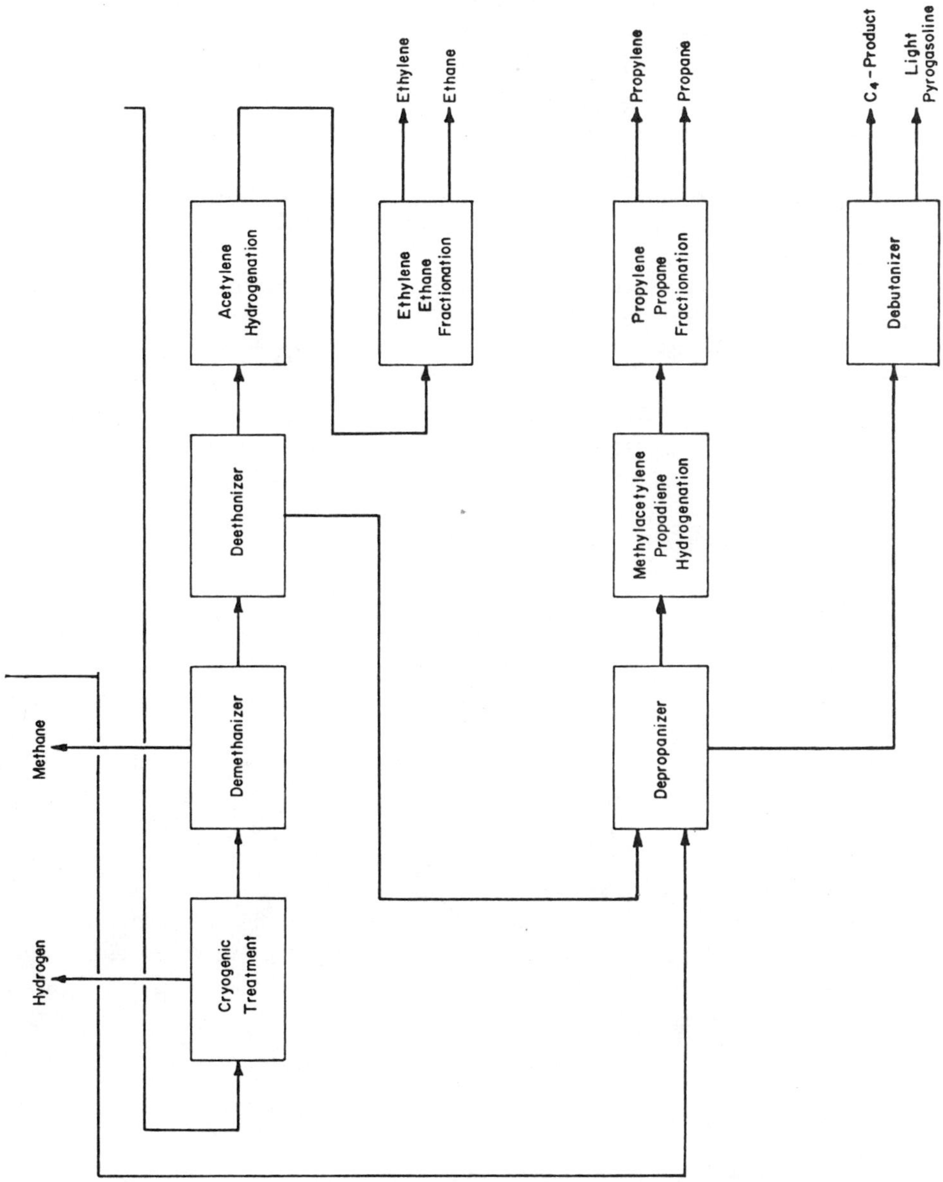

FIG. 24. Block flow diagram—ethylene production.

However, for further gas separation, systems have also been developed that call for the sharp separation of the C_4 and heavier components, or the C_3 and heavier components, from the pyrolysis gas as the first fractionation step. Although thermodynamic theory declares all paths to the pure products equivalent, most process designers appear to prefer demethanization as the first sharp fractionation step as is shown in the block flow diagram.

Several ethylene plants producing high-purity propylene do not contain facilities for the hydrogenation of methylacetylene and propadiene but instead fractionate the raw C_3-cut into high-purity propylene and a small fraction containing essentially all the propane, propadiene, and methylacetylene present in the pyrolysis gas plus sufficient propylene to maintain the concentration of the hazardous methylacetylene and propadiene well below established safety limits.

In recent years reduction of the energy requirements of ethylene plants has been a major objective of process development work in many companies. It is the manner of recovering waste heat, recuperating refrigeration, and avoiding thermodynamic inefficiencies that differentiates various modern ethylene processes. Sohns and Fuge [75] give a detailed analysis of the energy flow in an ethylene plant and discuss process options to reduce the energy losses.

7.2. A Typical Ethylene Manufacturing Process

Figures 25 through 30 show, in basic concept and without indication of the control logic, the process flow diagram of a typical plant producing high-purity ethylene and propylene from naphtha or gas oil. In fact, there is little fundamental process difference between plants designed to process naphtha and plants that handle gas oil. Minor differences exist in the design of the pyrolysis and waste heat recovery sections, and the different yield spectra call for commensurately different capacities of the separation systems. However, a plant designed to produce a given quantity of ethylene from gas oil will have essentially the same ethylene production capacity when processing naphtha. In general, a plant designed for one feedstock is always capable to process the next lighter (molecular weight or density) feedstock. The figures do not show the finer details of energy conservation which invariably involve greater complexity and integration of the distinct systems than can be shown in these simplified flow diagrams.

7.2.1. *Pyrolysis and Primary Waste Heat Recovery*

The pyrolysis section is considered the heart of an ethylene plant, this section producing all the products of the plant, while the other sections serve to separate and purify the products. Through the pyrolysis yield pattern, feedstock consumption, and efficiency of fuel utilization, the pyrolysis section has the greatest influence on the economics of the plant. For this reason pyrolysis and pyrolysis furnaces are treated in further detail in a subsequent chapter.

Depending on the plant and furnace capacity, the pyrolysis section of ethylene plants contains between four and 20 pyrolysis furnaces with an average of approximately 10 furnaces per plant. In most cases two furnaces are positioned with their convection sections back to back as shown in Fig. 25, sharing a common stack and sometimes an induced draft fan. Each furnace contains an even number of multiple pyrolysis coils in the radiant section. The exact number depends on the coil design concept of the furnace designer and the desired furnace capacity. Due to the high fire box temperature, more than 50% of the fired duty has to be absorbed in the convection section of the pyrolysis furnaces. The vaporization and preheating of the feedstock requires only a portion of the convection section duty. The remainder is utilized to preheat boiler feed water and to superheat high-pressure steam. The final steam superheat temperature is controlled through injection of boiler feed water, as is standard practice for high-pressure boilers.

The feedstocks are usually first preheated externally to the pyrolysis furnaces and then in the uppermost coil of the convection section. Liquid feedstocks are partially vaporized and then mixed with dilution steam in the weight ratio of between 0.5 to 1.0 steam to hydrocarbons. The dilution steam has the dual function of lowering the hydrocarbon partial pressure and reducing the carburization rate of the pyrolysis coils. The steam–hydrocarbon mixture is further heated in the convection section to the crossover temperature between the convection and radiant sections. This temperature has to be carefully selected for any given feedstock to obtain maximum heat absorption without reaching temperatures high enough to initiate cracking which could result in detrimental coke deposits in the convection section. Depending on the feedstock, this temperature is usually in the range of 550 to 700°C. Some furnace designers provide a bank of steam-cooled shock tubes for the lowest part of the convection section to protect the process convection coils.

The steam–hydrocarbon mixture then enters the pyrolysis coils where it is further heated and cracked into products. For the pyrolysis reactions, hydrocarbon partial pressure and temperature as functions of coil length and time are of particular importance. The total residence time the gas spends in the pyrolysis coil of a modern furnace is generally in the range of 0.2 to 0.6 s. The gas outlet temperature is in the range of 770 to 870°C depending upon feedstock and furnace design. The product gas is sent to closely connected transfer line exchangers in which the temperature is rapidly reduced by several hundred degrees by generating high-pressure saturated steam at approximately 12.5 MPa. Given the extreme process gas temperature and the high steam pressure, the transferline exchangers have to be of very special design and, in fact, their development over the last 15 years has contributed much to the efficiency of modern ethylene plants (see Section 8.6 for further details). Two or more transferline exchangers together with a common steam drum are usually built right into the furnace structure. One exchanger serves one or, more often, two pyrolysis coils. Boiler feedwater circulation from the steam drum through the transferline exchanger is by thermosyphon action with typical water-to-steam ratios of 10 or higher. In two Japanese designs [76, 77], each transferline exchanger has its own integral steam drum.

Pyrolysis furnaces and their transferline exchangers form a pyrolysis

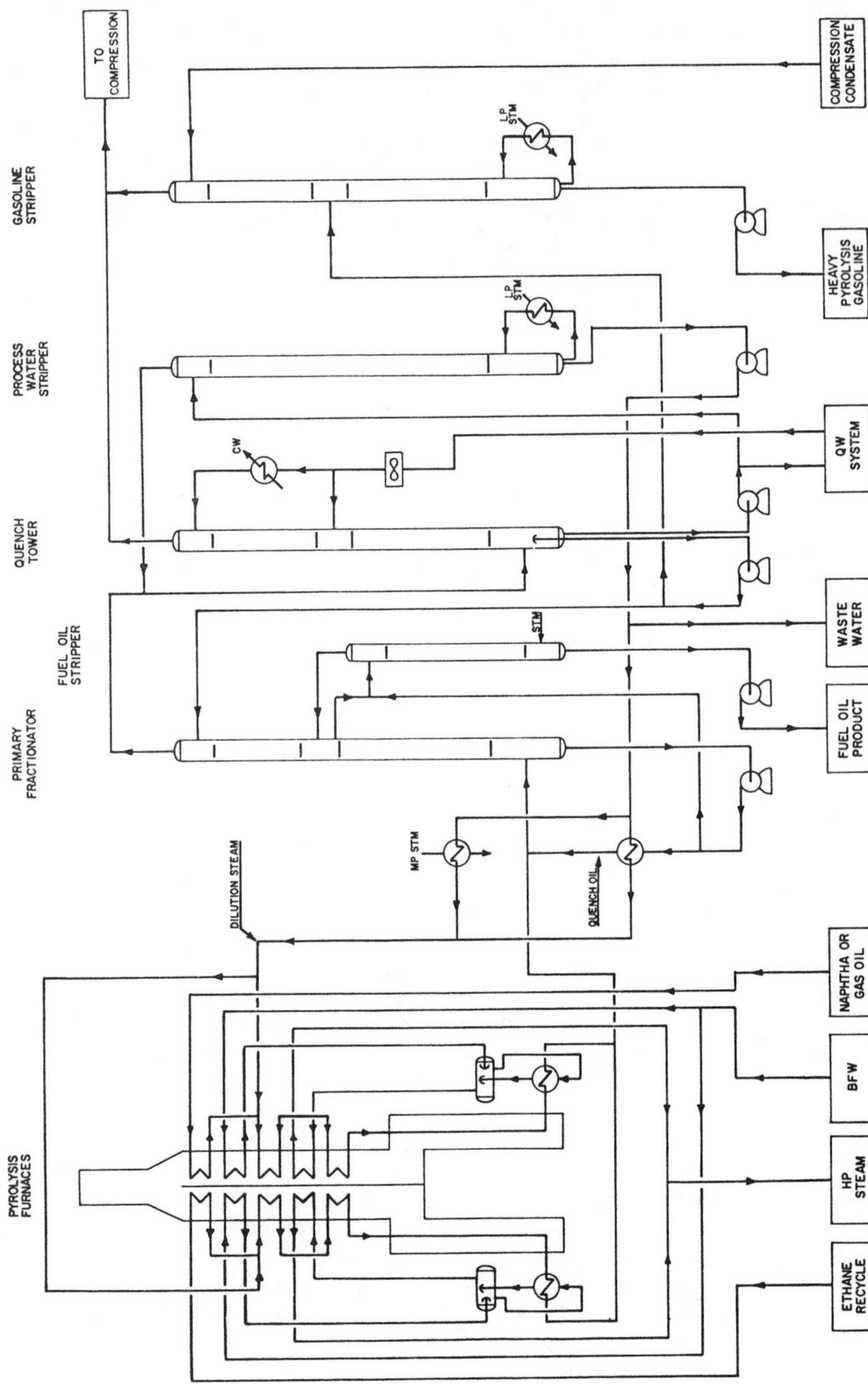

FIG. 25. Pyrolysis—waste heat recovery and heavy hydrocarbon separation.

module. The close integration is further enhanced by preheating of the boiler feedwater for, and the superheating of, the saturated steam from the transferline exchangers in the convection section of the pyrolysis furnace.

To a degree, the pyrolysis of all hydrocarbon feedstocks involves the complete dehydrogenation of some compounds which produces a slowly progressing deposition of coke on the internal surfaces of the pyrolysis coil. The coke tends to retard the heat transfer from the tubes to the gas, and this results in a slow increase of the tube metal temperature as the thickness of the coke layer increases. Depending on the diameter of the tubes of the coil, the coke deposit may also result in an appreciable increase of the pressure differential between inlet and outlet of the coil which has a detrimental effect on the pyrolysis yield pattern through an increase of the hydrocarbon partial pressure.

The pyrolysis gases of liquid feedstocks in particular contain components which tend to condense in the transferline exchanger and through further unknown reactions solidify on the internal surfaces of the tubes. Again the heat transfer rate is slowly reduced and the temperature of the gas leaving the transferline exchanger rises slowly. The principal reason for generating steam of such high pressure is to maintain the tube wall temperature as high as possible to minimize the condensation of tars.

The accumulation of deposits in the pyrolysis coils and transferline exchangers, which manifests itself by a rising coil inlet pressure (and hydrocarbon partial pressure) and rising metal temperatures, which eventually reach the maximum permissible service temperature of the materials, necessitates periodic shutdowns of the pyrolysis modules for the removal of these deposits. The designer will generally strive for conditions that result in equal times to reach given limits in the service temperature of the metals of the pyrolysis coil, the transferline exchanger, and the systems immediately downstream of the transferline exchanger.

Decoking of the pyrolysis coil is accomplished with steam or steam–air mixtures. With steam at temperatures of approximately 900 to 1000°C, the water gas reaction produces hydrogen and carbon monoxide while steam–air mixtures allow combustion of the coke at temperatures of 800 to 850°C. The deposits on the inner tube surfaces of the transferline exchangers are usually removed mechanically with a high-pressure water jet. For the two Japanese transferline exchangers mentioned earlier, regeneration through combustion is claimed as a major advantage.

Required spare heater capacity relates to pyrolysis coil and transferline exchanger run length and down time needed for decoking and cleaning. From the on-stream factor viewpoint, a practical approach for spare capacity in an ethylene unit heater section is about 10% [78].

7.2.2. Secondary Waste Heat Recovery and Pyrolysis Fuel Separation

The pyrolysis gas leaving the transferline exchangers is at temperatures that range from approximately 375 to 500°C in the case of naphtha pyrolysis and typically from 500 to 600°C in the case of gas oil pyrolysis. The outlet temperature depends on the amount of carbon deposits in the transferline exchangers. Since the rate of decomposition is higher at the beginning of an operation cycle, the average gas outlet temperature from all transferline exchangers is higher than the midpoint of the ranges above. In one process which is characterized by relatively high transferline exchanger outlet temperatures, depending on the pyrolysis feedstock a second transferline exchanger is employed to increase heat recovery further. Most processes also recover lower level heat through direct injection of quench oil into the pyrolysis gas either directly downstream of the individual transferline exchangers or into the transfer line immediately upstream of the primary fractionator. The first location is preferred for gas oil-based plants; the second location for naphtha-based plants. The first offers the advantage of reducing the pyrolysis gas temperature to approximately 200 to 220°C and thereby stopping most product degradation reactions. The transfer line can be made of normal carbon steel. It has the disadvantage that the transfer line must be designed for a very large mixed phase flow of pyrolysis gas and quench oil.

Pyrolysis gas and quench oil are separated in the base of the primary fractionator. The quench oil is cooled to approximately 185°C in the dilution steam generators or lower pressure steam generators (0.4 to 0.5 MPa) and in heat exchange with other process streams. The maximum temperature of the quench oil in the bottom of the primary fractionator has to be carefully controlled because of the instability of this material, certain components of which tend to polymerize with an attendant increase of the viscosity. The quench oil of gas oil-based plants exhibits higher thermal stability than does the material in naphtha-based plants. In the latter the material is entirely synthetic since the feedstock does not contain any components boiling in the pyrolysis fuel oil range. As the fuel oil yield from naphtha is significantly lower than from gas oil, the residence time of the quench oil in the system is proportionately longer, resulting in a greater accumulation of polymerization products. Quench oil often contains coke particles which can cause flow problems in the quench oil circuit.

Due to the small temperature difference between hot quench oil and the steam generators and the large amount of heat to be removed from the pyrolysis gas, the quench oil flow has to be large and typically is in the range of 15 to 25 times the flow of feedstock to the pyrolysis section. Three pumps including a spare are usually installed in the quench oil circuit which must be of special design because of the abrasive nature of coke particles contained in the quench oil.

The total heat transfer area of the quench oil-heated dilution steam generator is one of the largest of any service in an ethylene plant, with a total of 10 or more shells not being uncommon for a plant with a capacity of 500,000 t/yr ethylene. This total includes spares to permit mechanical removal of polymer deposits and coke during normal operation of the plant. Picciotti has analyzed the problem of optimizing the heat transfer area in quench oil circuits [79].

The function of the primary fractionator is to make a sharp separation between pyrolysis gasoline and pyrolysis fuel oil. This tower must assure the distillation end point of the pyrogasoline. Reflux is provided from gasoline condensed in the quench tower which, in effect, represents the reflux condenser of the primary fractionator. The overhead temperature is generally set to avoid the condensation of steam which, due to the presence of acidic components, could cause corrosion in the system. Since the bottoms temperature is selected for optimum heat recovery in the quench oil circuit, a sidestream withdrawal is provided as an exit for fuel oil components that are too volatile to be retained in the bottoms and too heavy for the pyrogasoline. The net fuel oil product is withdrawn from the bottom of the primary fractionator and sent to the fuel oil stripper. Gas oil-based ethylene plants often contain two strippers, one for the side stream and one for the heavy quench oil bottoms stream. The lighter material may be used as a fuel for hearth burners in the pyrolysis furnaces. In naphtha-based plants the total fuel production is so small that the separation into a distillate and a residual fuel oil is of little interest and therefore only one stripper is provided.

7.2.3. Tertiary Waste Heat Recovery and Heavy Gasoline Separation

The gas leaving the primary fractionator has a temperature of between 100 and 110°C and is further cooled in air- or water-cooled heat exchangers or in a direct contact cooled quench tower. The latter has the inherent advantage of low-pressure drop and recovers the residual heat of the pyrolysis gas through absorption in hot quench water. This hot quench water is usually available at a temperature of about 80°C for various low-level heating services throughout the plant. Since most of the latent heat of the dilution steam is released in the lower section of the quench tower, the bulk of moderately cooled quench water is added to that section, while a smaller stream is cooled as low as possible and fed to the top section of the tower to lower the temperature of the pyrolysis gas as much as possible before sending it to the main compressor. A detailed analysis and design procedures for the quench tower were given by Picciotti [80].

In the base of the tower, hot quench water is separated from heavy pyrogasoline which is condensed along with the dilution steam. The bulk of the hydrocarbons is pumped to the primary fractionator to serve as reflux, while the net product is sent to the heavy pyrogasoline stripper for the removal of C_4 and lighter hydrocarbons.

The hot quench water is sent to various heat exchangers and, in particular, the reboiler of the propylene fractionator if the plant produces high-purity propylene. Residual heat of the quench water is rejected to the atmosphere in a combination of air and water coolers.

The net dilution steam condensed is withdrawn from the quench water and, after a second more effective water–hydrocarbon phase separation, is sent to the process water stripper for the removal of dissolved hydrocarbons and gases. The treated process water is then pumped to the dilution steam generators which are heated with quench oil and medium-pressure steam (1.2 to 1.5 MPa). The net process water discharge from the system is minimal since only the small quantity of live stripping steam fed to the fuel oil strippers represents a net inflow.

The quench water, due to its absorption of acidic compounds, is potentially corrosive. Various neutralizing agents and other types of corrosion inhibitors are added to the quench water to protect the equipment. Careful control of pH value is important as high pH values have been found to lead to foaming and the formation of emulsions which are difficult to break.

7.2.4. Pyrolysis Gas Compression, Acid Gas Removal, and Drying (Fig. 26)

The pyrolysis gas leaving the quench tower usually has a temperature of 35 to 40°C and a pressure slightly above atmospheric pressure. Most processes call for compression to a pressure of approximately 3.5 MPa which appears optimal for the subsequent cryogenic treatment. The pyrolysis gas, in particular, if derived from liquid feedstocks, contains appreciable quantities of highly reactive diolefins and acidic components. The diolefins tend to polymerize at elevated temperatures and pose the danger of polymer deposits in the compression system. In fact, polymer formation is of concern throughout the entire process wherever elevated temperatures and moderate to high concentrations of diolefins exist. The acidic components raise the spectre of stress corrosion.

Modern ethylene plants employ radial centrifugal compressors exclusively; the future may possibly see the introduction of more efficient axial compressors for the early stages of pyrolysis gas compression in very large capacity plants. Most operators and designers of ethylene plants limit the maximum discharge temperature from the individual stages of the pyrolysis gas compressor to minimize the danger to polymer deposits. The temperature limits largely determine the number of compression stages. Plants based on gaseous feedstocks generally employ four stages while many naphtha- and gas oil-based plants employ five stages of pyrolysis gas compression. However, even a five-stage compression system will not yield discharge temperatures as low as 85°C with typical cooling water temperatures available at most plant locations. If lower compressor discharge temperatures are desired, cooling with water followed by high-level propylene refrigeration is required which then permits the use of a four-stage pyrolysis gas compressor. Although the five-stage pyrolysis gas compression system is more common, four-stage systems are in operation with and without propylene intercooling.

The typical pyrolysis gas compressor consists of three compressor casings driven by a single extraction/condensing steam turbine. For very large plants, two moderate size turbines are sometimes preferred, one driving two compressor casings and the second turbine the remaining one. The maximum allowable stress in individual compressor wheels is normally limited to 620 MPa if stress corrosion is of concern.

Water and hydrocarbons condensed between stages are separated from the pyrolysis gas in interstage separators. The water is returned to the quench water system while the hydrocarbons formed in the first three stages is sent to the heavy pyrogasoline stripper while the condensate from the last one or two stages is sent to the condensate deethanizer.

Hydrogen sulfide and carbon dioxide are removed from the pyrolysis gas

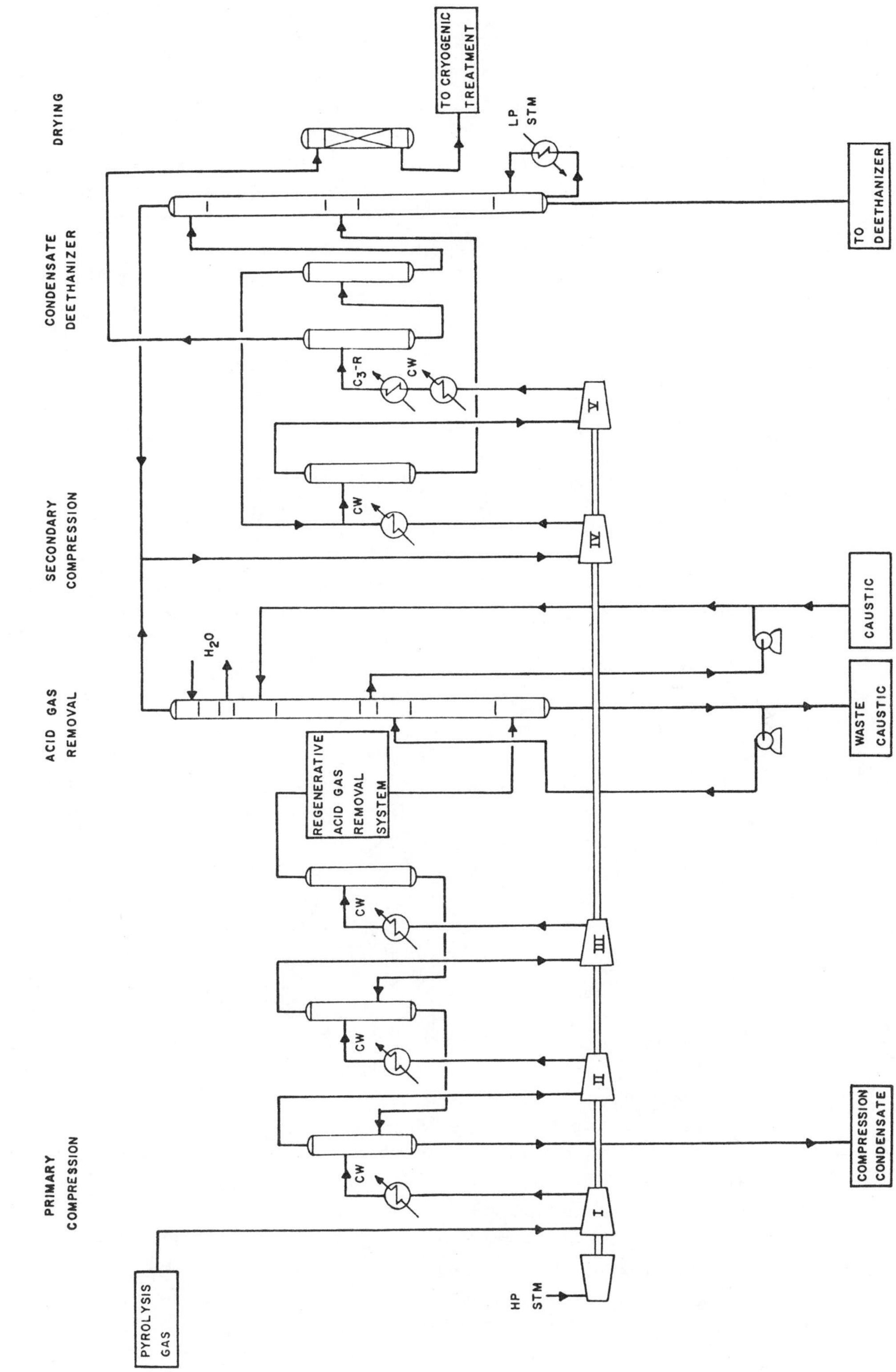

FIG. 26. Pyrolysis gas conditioning.

between the third and fourth stages of the compression system. This location is optimum as the actual gas volume has been reduced significantly in the first three stages of compression while the acidic components are still present in the gas stream and have not contaminated any products that are separated ahead.

In those plants processing feedstocks with a sulfur content of a few hundred ppm, scrubbing with a dilute caustic soda solution has proven most economic. A two- or sometimes three-stage scrubbing tower is generally provided to obtain maximum utilization of the caustic soda. The solution circulating over the top section of the scrubbing tower typically contains between 8 and 12% free sodium hydroxide while the solution in the lower section will contain between 2 and 5% free caustic. A detailed analysis of the number of stages, number of contacting plates, and strength of the caustic solution has been published by Raab [81]. Relatively weak caustic solutions are preferred to avoid the precipitation of sodium salts and to minimize the formation of organic sodium complexes. The pyrolysis gas leaving the caustic scrubber contains less than 1 ppm of acid gases and thus assures that the final products of the plant will meet specifications in this respect. The spent caustic represents environmentally the most troublesome liquid effluent of ethylene plants and requires substantial subsequent treatment before it can be safely discharged.

Plants designed to process feedstocks with significant sulfur contents (i.e., in excess of 500 ppm approximately) often contain a regenerative acid gas removal system upstream of the caustic scrubber. Regenerative systems alone cannot achieve the essentially complete removal of acid gases from the pyrolysis gas as required. These systems, which may employ monoethanolamine, diethanolamine, or Alkazid as solvents, would be of the standard absorber-desorber design if it were not for the presence of diolefinic compounds in the pyrolysis gas which tend to dissolve in, or to be weakly bound to the solution and, through polymerization, may cause severe operational problems in the desorber. These systems must contain facilities for the removal of hydrocarbons from the rich solution. In fact, it was the development of these solution purification techniques that, after unsatisfactory early attempts, made possible the application of regenerative acid gas removal systems for the bulk removal of hydrogen sulfide and carbon dioxide from pyrolysis gas.

Depending on the environmental requirements at specific plant locations, the acid gases leaving the regenerative acid gas removal and spent caustic neutralization systems are either burned and the combustion products sent to flue gas, or, as is becoming more common, are treated in a Claus unit for conversion of the hydrogen sulfide to elemental sulfur.

Complete removal of water from the pyrolysis gas is required in preparation for the cryogenic treatment. One process employs a dual drying system consisting of bulk water removal facilities employing glycol immediately downstream of the acid gas removal system and of conventional adsorption on solid desiccant after final compression and some precooling of the pyrolysis gas. This system has the advantage of eliminating water from the condensates in the final stages of compression and the subsequent condensate processing facilities. Most processes, however, employ a single adsorptive drying system located immediately after the final stage of pyrolysis gas compression. Molecular sieves are the preferred desiccants because of their selectivity. Of two dryers normally provided, one is in normal operation while the other is regenerated, recooled, or held in a standby position. High-pressure methane heated with steam or in a direct-fired heater to a temperature of approximately 225°C is the preferred regeneration medium. The dryers are designed for on-stream times of 24 to 48 h between successive regenerations.

To limit the quantity of desiccant required and thus the size of the vessels, the pyrolysis gas is, upstream of the dryers, cooled with propylene refrigeration to above the hydrate inception temperature, the value of which depends on the system pressure.

Many older plants use activated alumina in a three-dryer arrangement of which two are in operation in series while the third is in regeneration. Activated alumina is less selective than molecular sieves and adsorbs, along with water, some unsaturated hydrocarbons. Through complete saturation with water of the desiccant in the first dryer, the adsorbed hydrocarbons can be desorbed to some degree although not sufficiently to avoid all loss to the regeneration gas or to prevent the slow deactivation of the desiccant through deposition of polymers on its active surface.

7.2.5. *Cryogenic Treatment and Demethanizer*

The cryogenic treatment involves the successive chilling of the pyrolysis gas in heat exchange with boiling refrigerant and with cold process streams that have to be vaporized and/or reheated. As the pyrolysis gas temperature is reduced, more and more condensate is formed. There exist numerous process variations on the sequence of condensate separation, fractionation, and heat exchange.

In one process the pyrolysis gas is, after moderate cooling and some condensation, fractionated sharply into a C_3 and heavier hydrocarbon fraction and a C_2 and lighter fraction. The former is sent to the depropanizer, the latter is removed as a vapor stream and reheated for hydrogenation of the acetylene and again cooled for further condensation and separation into hydrogen, methane, and an ethylene–ethane fraction.

In another process the entire pyrolysis gas is, after some cooling and condensation, sent to a high-pressure demethanizer which accomplishes the essentially complete removal of methane from the condensate but loses some ethylene in the hydrogen–methane overhead product. Ethylene refrigerant boiling in the reflux condenser at slightly above atmospheric pressure controls the dew point temperature of the overhead product. The ethylene contained in the overhead product is usually recovered either through condensation and recycling to the main compressor or through condensation with refrigeration generated by a turboexpander.

Most processes, even those designed for the separation of C_4 and heavier hydrocarbons in the compression system, contain an arrangement similar to that shown in Fig. 27 in which only the condensates are sent to the demethanizer. Operating at a pressure of approximately 3.2 MPa, the tower is designed for essentially complete separation of methane from ethylene and heavier components. Condensing propylene refrigerant supplies the necessary heat to the reboiler, and vaporizing ethylene condenses reflux and some of the overhead product. The overhead product consists mostly of methane with a small amount of hydrogen which enters the tower dissolved in the condensates.

Most modern ethylene plants also produce a high-purity hydrogen stream (95 mol% being typical) and a concentrated methane-rich off-gas stream (95 mol% methane with the remainder mostly hydrogen). In those processes from which successively colder liquid streams are condensed from the dried and compressed pyrolysis gas, the hydrogen stream is withdrawn from the cryogenic treatment section upstream of the demethanizer tower as shown in Fig. 27a. In those processes in which the entire pyrolysis gas stream is sent to the demethanizer, the hydrogen stream is withdrawn downstream of the demethanizer as shown in Fig. 27b.

The cryogenic treatment systems of all ethylene processes contain inherent thermodynamic irreversibilities such as the isenthalpic throttling of the demethanizer overhead and fractionators with large temperature differences between top and bottom, more than 100°C in the case of the high-pressure demethanizer. Side vaporizers for fractionators [82, 83] can significantly reduce the irreversibilities of fractionation systems and are, in fact, applied in ethylene plants to save energy. Interest in a three-stage propylene–ethylene–methane refrigeration cascade, which was used in early small-scale ethylene plants [84], has recently revived. Provision of methane refrigeration allows the design of an efficient low-pressure demethanization system but requires the reintroduction of reciprocating compressors. Lam et al. [85] report on the use of a low-pressure demethanizer.

The practice of using multicomponent refrigerants for natural gas liquefaction plants has led to an investigation of the suitability of this concept for the cryogenic treatment of pyrolysis gas. The condensates formed can be used as multicomponent refrigerants with the main compressor also serving as the refrigerant compressor.

Brazed aluminum plate-fin exchangers are generally used for the multipass cryogenic heat transfer services and are sometimes even used for the refrigerated chillers with thermosyphon circuits for refrigerant circulation. Much of the cryogenic equipment is often installed in a rectangular carbon steel container, commonly called cold box, with its void spaces filled with perlite or rockwool for insulation. Considering the complexity of the system, this mechanical design is characterized by surprising compactness and minimum space requirements.

7.2.6. Deethanizer and Acetylene Hydrogenation

The deethanizer is a conventional tray-type fractionator operating at a pressure of 2.4 to 2.8 MPa to separate the demethanizer bottoms product from an overhead stream containing all three C_2-hydrocarbons and a bottoms product of C_3 and heavier hydrocarbons. Essentially complete removal of C_2-hydrocarbons from the bottoms product is more important than the purity of the overhead product. Any C_2-hydrocarbons contained in the bottoms product will eventually reach the propylene product which often has a 10-ppm

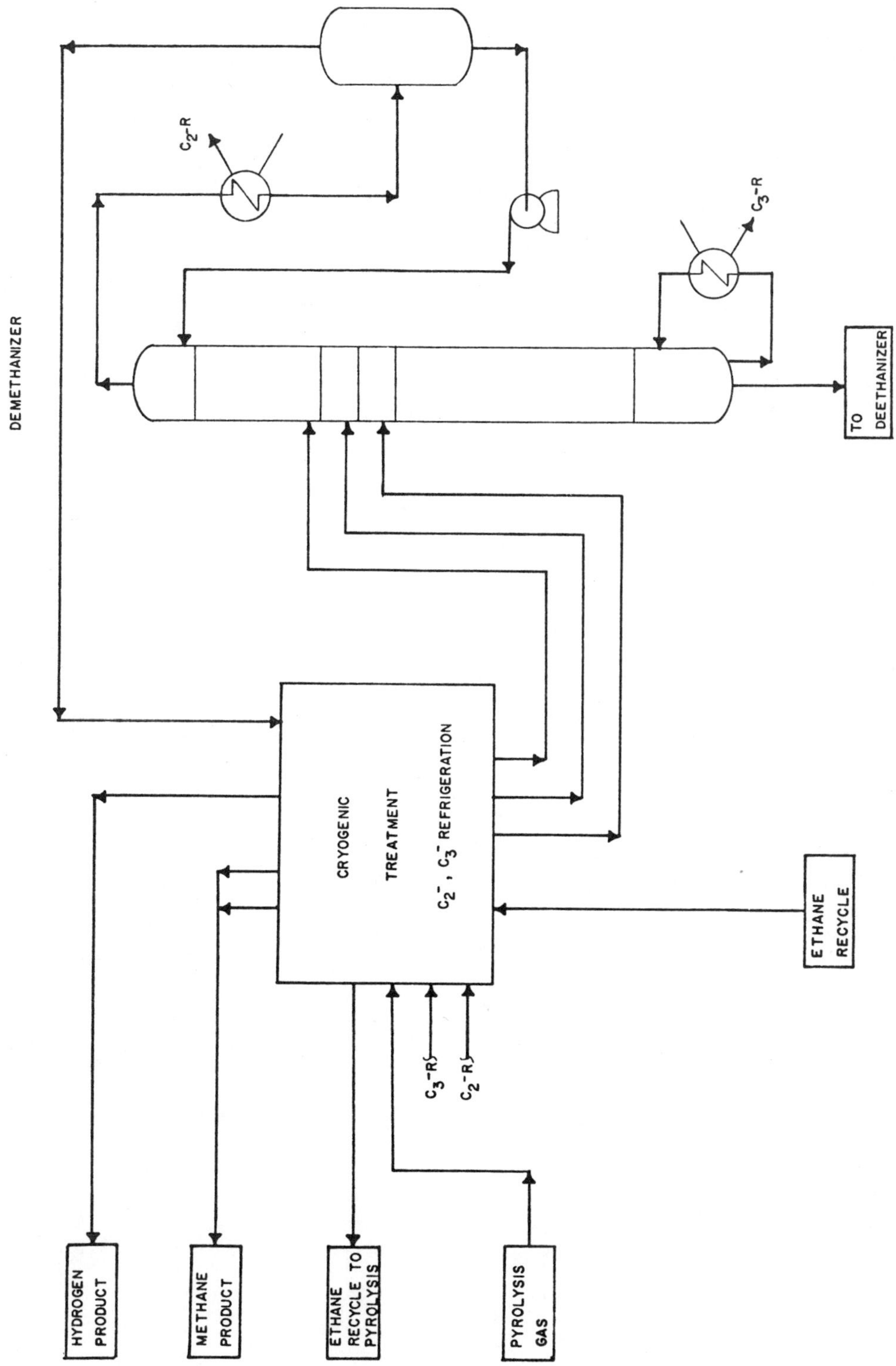

FIG. 27a. Cryogenic treatment and demethanizer.

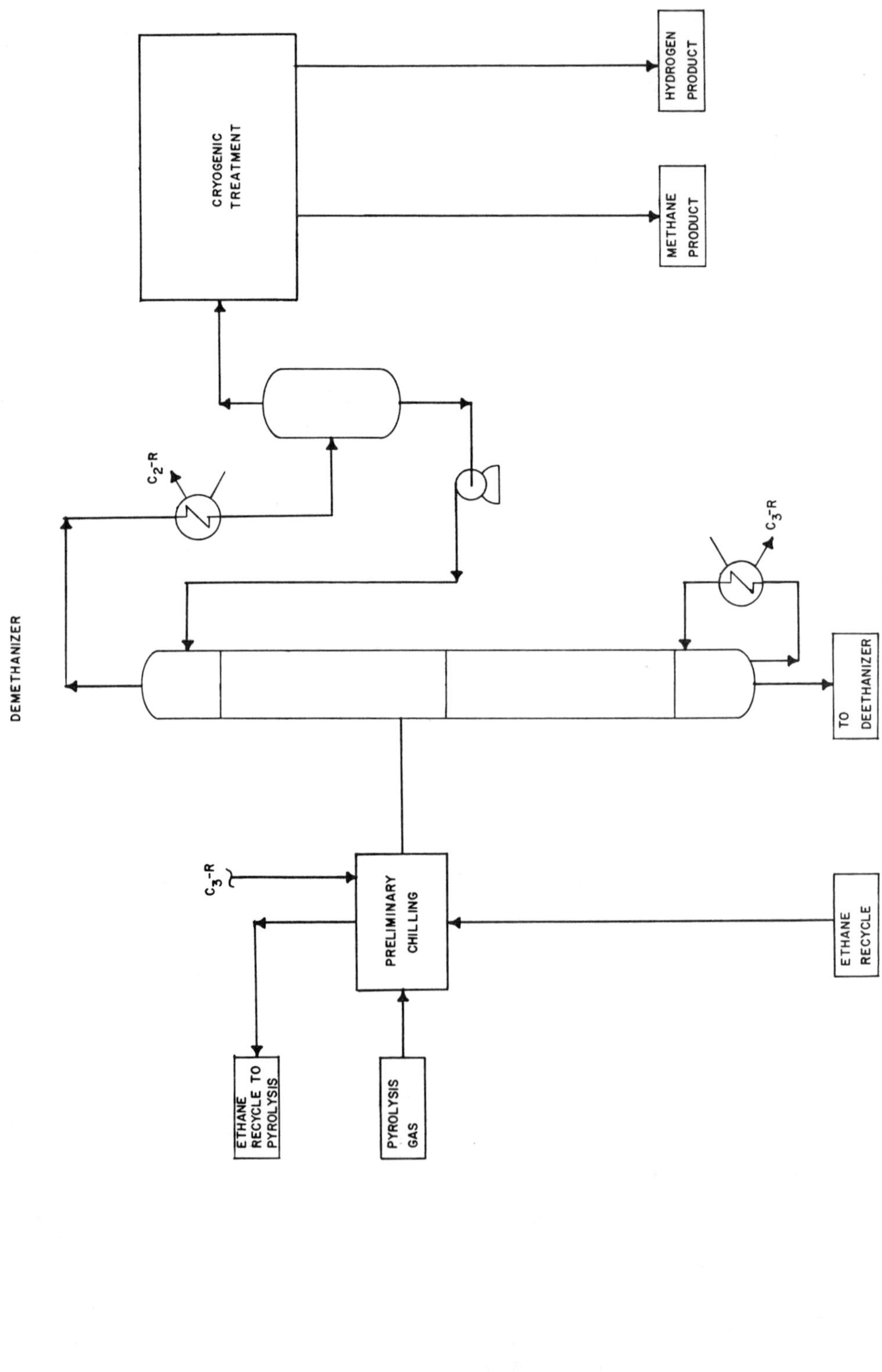

FIG. 27b. Cryogenic treatment and demethanizer.

specification on the ethylene content. Minor amounts of C_3-hydrocarbons in the overhead will be concentrated in the ethane product from the ethylene fractionator which is either recycled to pyrolysis or sent to fuel with neither utilization being sensitive to minor contamination with C_3-hydrocarbons. Low-pressure steam heats the reboiler and propylene refrigerant condenses the reflux.

Treatment of the C_2-stream can take various routes, depending on the acetylene content and on whether or not high-purity acetylene is to be recovered. The system shown in Fig. 28 is suitable for moderate acetylene contents when the recovery of acetylene is not of interest. The deethanizer gross overhead is heated to a temperature of between 20 and 100°C, hydrogen is added in the molar ratio of approximately 2 referred to acetylene, and the mixture is passed over a fixed bed of palladium catalyst. The heat of reaction is removed in a water cooler and the reactor effluent is further cooled by heat exchange with the feed and partially condensed by heat exchange with propylene refrigerant. The liquid is pumped as reflux to the deethanizer while the vapor flows through the guard dryer to the ethylene–ethane fractionator. The reactor effluent typically contains less than 1 ppm of acetylene but is contaminated with traces of hydrogen and methane which represent the major disadvantage of this location of the acetylene hydrogenation system. Yet this location has the advantage over the front end hydrogenation in that it allows very accurate control of the hydrogen concentration and reaction temperature, resulting in good selectivity of the reaction. Under normal conditions as much acetylene is hydrogenated to ethylene as is ethylene to ethane with no net loss of ethylene. It has been found that traces of carbon monoxide increase the selectivity of the catalyst.

Some undesirable polymerization side reactions occur in the hydrogenation reactors, resulting in the formation of olefins with carbon numbers ranging from four to over 20. The side reactions in the hydrogenation reactor cause a slow deactivation of the catalyst which calls for gradually rising inlet temperatures to maintain adequate activity. Where lower acetylene concentrations are present in the C_2 fraction, the overhead product alone is usually hydrogenated. Often two stages are designed into a single reactor with water cooling between stages and separate hydrogen addition controls for each stage. Systems consisting of an internally cooled isothermal reactor followed by an adiabatic reactor have been employed successfully but have not found widespread acceptance. Removal of the polymers and residual hydrogen and methane is sometimes accomplished in a two-tower fractionation system. Ultimately, after several months of operation, regeneration of the catalyst is required. Controlled combustion of the deposits with a steam–air mixture at about 750°C completely restores the activity of the catalyst.

Acetylene recovery facilities can be located between the deethanizer and the ethylene–ethane fractionator or downstream of the latter. The typical acetylene recovery process makes use of three towers, the first to absorb acetylene in a suitable solvent, the second to reject and return coabsorbed ethylene and ethane from the solution, and the third to desorb acetylene [86, 87]. Acetone, dimethylformamide, and *n*-methyl pyrolidone are the preferred solvents. Certain C_3-hydrocarbons in the deethanizer overhead have solubilities in these

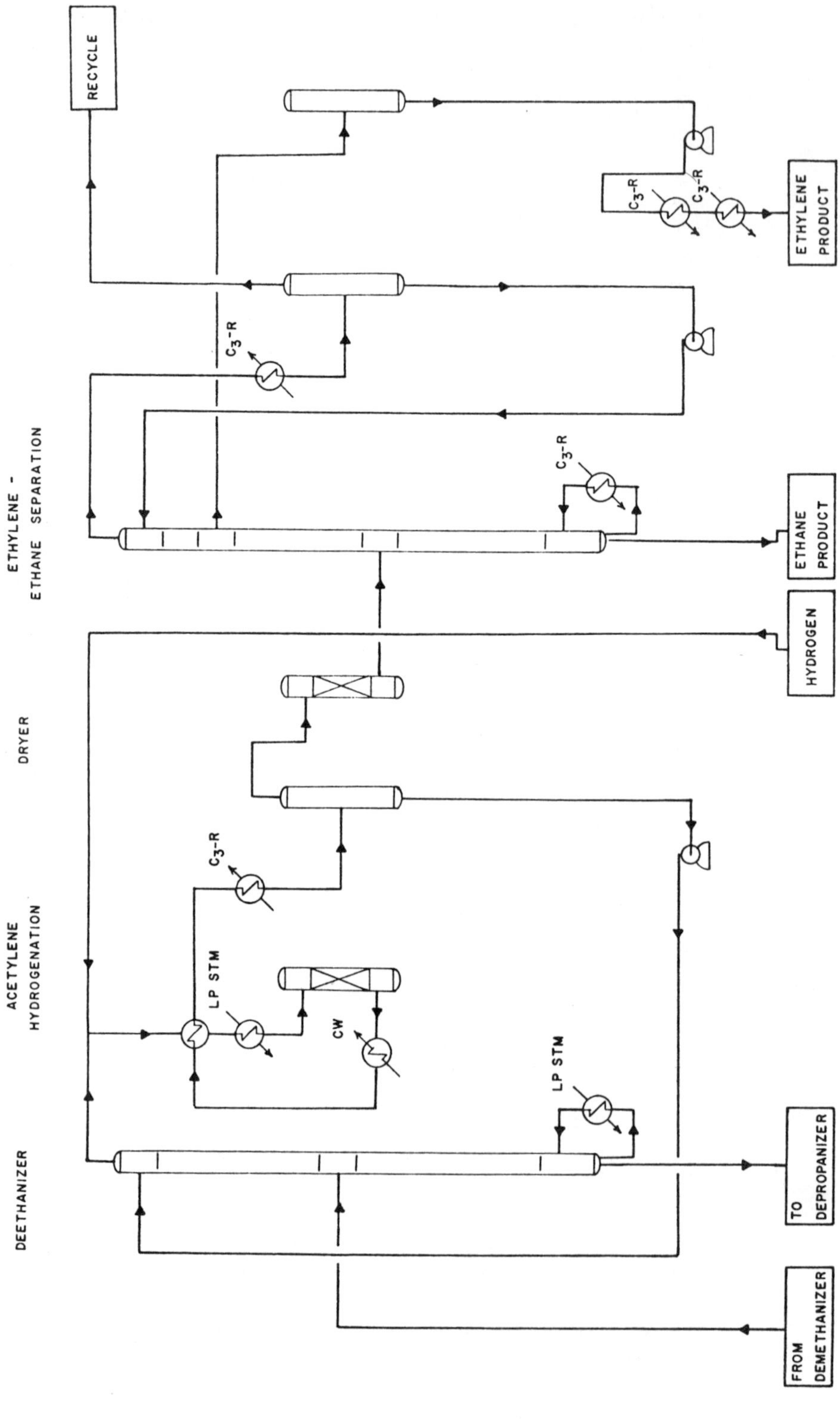

FIG. 28. Deethanizer, acetylene hydrogenation, and ethylene–ethane separation.

solvents comparable to the solubility of acetylene and therefore have to be completely removed from the feed. Location of acetylene recovery facilities downstream of the main ethylene–ethane fractionators eliminates this problem but involves higher concentrations of acetylene in this tower, resulting in partial pressures that, however, remain well below established safety limits. Acetylene recovery systems in both locations are presently in operation with a better overall safety record than acetylene hydrogenation systems have had.

7.2.7. Ethylene–Ethane Fractionation

The ethylene–ethane mixture leaving the hydrogenation system may contain traces of hydrogen and methane. It will contain all the acetylene of the pyrolysis gas if the process provides for acetylene recovery from the overhead product of the ethylene–ethane fractionator. In the first case with preceding acetylene hydrogenation the ethylene–ethane separation represents a classical binary fractionation problem while in the second case the separation concerns a ternary system. In the latter, acetylene is fractionated into the overhead but exhibits a concentration maximum within the tower. This concentration maximum must not exceed certain safety limits which are dependent on the operating pressure.

In either case the typical ethylene–ethane fractionator contains between 80 and 120 trays and requires a reflux ratio of between 2.5 and 4.0 depending on the operating pressure and product specification. A side vaporizer, which implies a savings in energy, is often applied to provide a significant portion of the total reboiler heat at a lower temperature level than the main reboiler.

As shown in Fig. 28, a "pasteurizing" section is occasionally provided in the top of the fractionator for the removal of hydrogen, carbon monoxide, and methane. This process element permits these contaminants to be reduced by factors approximately equal to the equilibrium constants of the contaminants under conditions as they exist at the point of ethylene withdrawal. A small stream containing most of the contaminants but still rich in ethylene is reheated and recycled to the main compressor.

The conditions of the reboiler and overhead condenser are perfect for the application of a heat pump. Most processes use a closed heat pump with propylene refrigerant as the working fluid. Condensing refrigerant vapor supplies heat to the reboiler, and refrigerant boiling under a lower pressure generates the refrigeration for the overhead condenser with the main propylene refrigerant compressor providing the necessary mechanical work. An open heat pump can be used with integration of the ethylene–ethane fractionator into the ethylene refrigeration system. Condensing ethylene refrigerant supplies the necessary heat for the reboiler of the fractionator, while liquid ethylene refrigerant is used directly as reflux. The fractionator pressure has to be lower with the open heat pump to avoid unduly high discharge pressures on the ethylene refrigerant compressor. The open as compared to the closed heat pump has lower energy requirements and eliminates the large overhead reflux condenser, but poses the danger of ethylene product contamination through minute leaks in any of the exchangers with ethylene refrigerant at a lower pressure than the process streams. The open heat pump with ethylene also has the advantage that propylene and ethylene refrigerant compressors have comparable power consumption. With the closed propylene heat pump the propylene refrigerant compressor consumes approximately five times as much energy as the ethylene refrigerant compressor.

7.2.8. *Purification of Propylene and Heavier Products*

The bottoms products of the condensate stripper and deethanizer are processed in the depropanizer for a sharp separation of the C_3-hydrocarbons from C_4 and heavier hydrocarbons. The bottoms product of this tower has a high content of butadiene and heavier diolefins which tend to polymerize at moderate temperatures. Deposits of a rubberlike polymer on the inner surfaces of the reboiler tubes and the internals of the tower can severely curtail the capacity of this fractionation system. Mechanical cleaning may thus be required every few months.

Two designs have been developed to overcome this problem. In the first the bottoms temperature is set low, which then results in an operating pressure for the system that requires propylene refrigerant instead of cooling water for the condensation of the overhead product. A full capacity spare reboiler is normally provided because the gross bottoms is still exposed to moderately high steam temperatures although the spare is only rarely used. In the second design the reflux is condensed with cooling water which calls for higher pressures and consequently for higher bottoms temperatures which will be excessive. To avoid a complete shutdown for the mechanical cleaning operation, the entire bottom section of the tower, including the reboiler, is spared.

The depropanizer bottoms is further processed in the debutanizer for separation of C_4-product from the light pyrogasoline. The debutanizer, operating at a moderate pressure of 0.4 to 0.5 MPa, is a conventional fractionator with a steam-heated reboiler and a water-cooled condenser. A spare reboiler is often provided although deposition of polymers is much less of a problem in this tower than in the depropanizer.

The overhead product of the depropanizer can be sent directly to the propylene–propane fractionator. However, it will require hydrogenation of the contained methylacetylene and propadiene if the bottoms product of the fractionator is to meet LPG specifications or if only so-called chemical-grade propylene is to be produced. The latter requires only stripping for removal of hydrogen and methane, and no fractionation of the hydrotreated depropanizer overhead.

Treatment of the depropanizer overhead for hydrogenation of the methylacetylene and propadiene over a palladium catalyst as shown in Fig. 29 is typically carried out in a vapor-phase system quite similar to the acetylene hydrogenation system described earlier. The hydrogenation can also be accomplished in proprietary mixed-phase isothermal systems.

Due to the low relative volatility of propylene to propane, the fractionation of propylene and propane is even more difficult than the fractionation of

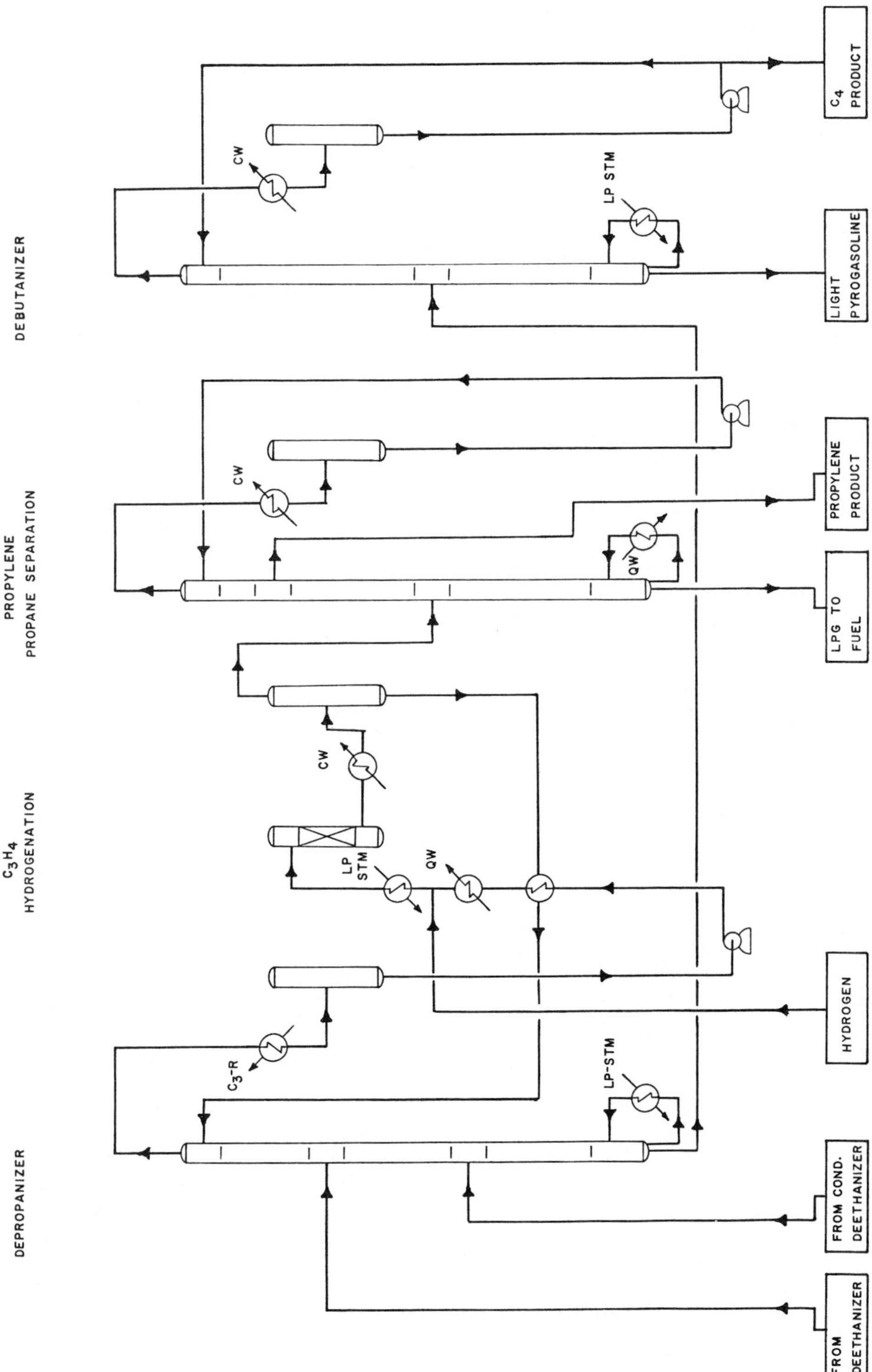

FIG. 29. Depropanizer, propylene fractionation, and debutanizer.

ethylene and ethane. Yet with the low-level heat available in the quench water of the tertiary heat recovery system of an ethylene plant, this separation can be accomplished without any supply of mechanical energy other than that required for the reflux pumps and possibly the liquid transfer pumps if more than one tower is required to accommodate the large number of trays. The typical propylene–propane fractionation system operates at a pressure of 1.8 to 2.0 MPa and requires 160 and more fractionating trays, with the exact number dependent on the purity of the propylene desired and the system pressure.

As the depropanizer overhead product normally contains well in excess of 90% propylene, most propylene–propane fractionation systems are designed for a propylene content of 10 to 40% in the propane product to keep the number of fractionating trays in a practical range. Design for lower propylene concentration of the propane product will add only little to the recovery of high-purity propylene.

In fractionation systems which process a depropanizer net overhead product directly, i.e., without hydrogenation of the contained methylacetylene and propadiene, the concentration of these components in the fractionator is cause for concern because, like acetylene, they can decompose into carbon and hydrogen with a destructive release of energy. For this reason the maximum concentration of these components has been set at 40%, and this has proven to be entirely safe by many years of completely accident-free operation in a good number of plants. In respect to safety, the relatively low temperature of the quench water of 80 to 85°C appears to be a real advantage as experiments have indicated that the decomposition of methylacetylene and propadiene in even higher concentrations requires high-energy spark initiation [88].

As in the ethylene–ethane fractionation, a pasteurizing section is occasionally incorporated into the propylene–propane fractionator for the rejection of hydrogen, carbon monoxide, and methane.

7.2.9. Refrigeration System

The separation of pyrolysis gas through condensation and fractionation at cryogenic temperatures requires external refrigeration over the entire temperature range from ambient to −100°C. Still lower temperatures, reaching as low as −165°C, can be generated using process streams.

The refrigeration system of the typical ethylene plant employs propylene and ethylene as refrigerants because they are suitable and readily available in the plant. The propylene refrigeration system has either three or four stages while the ethylene system has two or three stages for a total of between five and seven stages for the entire refrigeration cascade. Systems with a greater number of stages have a higher efficiency but require more piping and exchangers.

A typical seven-stage propylene–ethylene refrigeration cascade is shown in Fig. 30. Propylene refrigerant is not only used to supply refrigeration to chillers and condensers but also provides low-level heat to the top row of exchangers in Fig. 30, thereby, in effect, recuperating refrigeration. The reboilers of the demethanizer and the ethylene–ethane fractionator are opportune examples

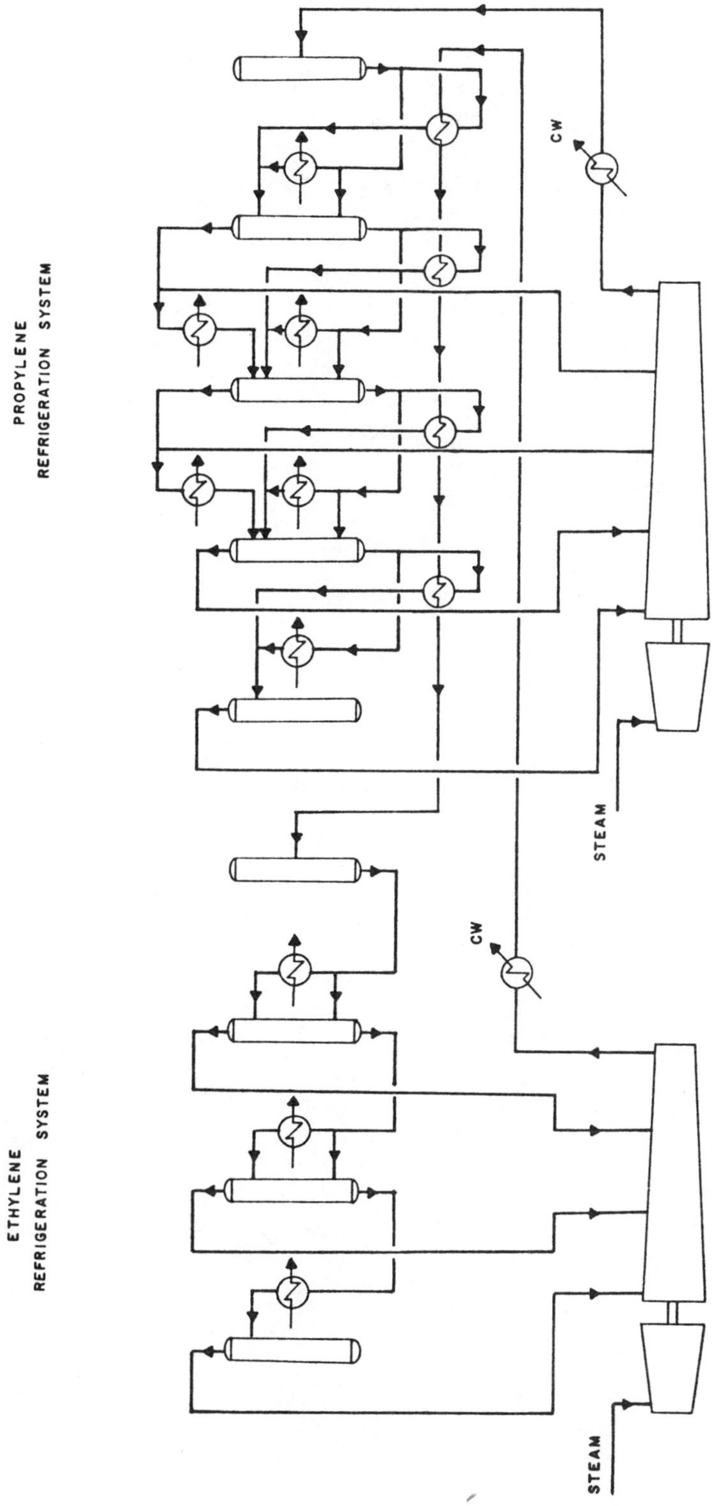

FIG. 30. Refrigeration cascade.

for refrigeration recuperation. In one process, ethylene refrigerant also recuperates refrigeration.

The vapors of the individual stages flow to the respective stages of the refrigerant compressors. Depending on the low-level heat requirements of a particular process, propylene may sometimes be flowing from, rather than to, the higher stages of the compressor. Propylene refrigerant is condensed against cooling water while ethylene refrigerant is desuperheated against cooling water and higher level propylene refrigerant, being finally condensed with low-level propylene refrigerant. Single centrifugal compressor casings can accommodate the respective number of stages of the ethylene and propylene system. The combined shaft power requirement of the two compressors is typically in the same general range as the power requirement of the pyrolysis gas compressor. The compressors are normally driven by steam turbines which may be of the condensing, backpressure, or condensing/extraction type as the plant steam balance requires.

Other refrigerants than propylene, notably propane and ammonia, have on occasion been employed, but there is no attractive alternative for the ethylene refrigerant. However, as indicated earlier, multicomponent refrigerants of the type used in LNG plants might be a possible alternative to the refrigeration cascade.

7.3. The Role of the Computer

In the 1950s engineers used slide rules and possibly electromechanical calculators together with simple and less accurate chemical and physical data systems to design complex ethylene plants. Twenty years later the design of a modern plant would be inconceivable without the use of large-scale computers. The electronic computer and the development of a multitude of sophisticated programs have contributed much to the efficiency that modern ethylene plants have attained. The computer plays an important role in all phases of constructing an ethylene plant, from process design through start-up.

General simulation programs have been developed for the calculation of mass and energy balances of process plants. These programs are normally of the modular type which allow the process designer to compose a model of his process that he can modify at will to study the effect of variations. Modular simulation programs are most useful in the design of process plants. Simpler nonmodular simulation programs representing the model of a specific process often contain an optimization algorithm and are useful for the owner to optimize the performance of a given plant with different feedstocks and varying production objectives.

A generalized modular process simulator determines the mass and energy balance of a process unit under steady-state conditions. Each module type is implemented in a computer subprogram which performs a fixed type of calculation. All such modules represent models of classical unit operations such as phase separators, heat exchangers, fractionators, and reactors which can be arranged and interconnected as required by the process flow scheme to form a mathematical model of an entire process or a process section. A simulation problem requires, as input, definition of feeds, the types of modules and their manner of interconnection, plus a number of fixed operating parameters to make the system mathematically determinate. The program then performs highly complex iterative calculations leading to a unique solution if such, in fact, exists for the problem as defined. Thus, even complex flow schemes, often involving recycle streams as do ethylene plants, can be solved with speed and mathematical accuracy.

An essential part of any generalized simulation program is the data package. While it is possible to supply, for each problem, the vapor–liquid equilibrium and enthalpy data in the form of tables, general data correlations are more convenient and more accurate. Most simulators have generalized physical data systems built into the computer system.

A computerized package is made up of two parts: a data bank containing the necessary constants for each compound and general or specific correlations for calculating the required physical properties. A typical data bank covers a reasonable number (100 to 300) of compounds which are considered important to the most common set of problems. However, this basic number of compounds is augmented and considerable flexibility is attained by allowing for the ad hoc addition of compounds to the data bank by supplying the necessary constants as part of the input to the program. This feature not only allows a wide variety of compounds to be handled by the program, but also offers the option of tailor-made correlations to suit special problems. Narrow boiling petroleum fractions are handled as a single special type of compound. For each fraction the mean average boiling point and the specific gravity are used as characteristics. Through built-in correlations for such important quantities as molecular weight and critical properties, the necessary basic constants are generated so that petroleum fractions can be handled by the program in the same way as other single compounds.

To date, no single correlation has been developed to provide accurate data for all thermodynamic problems, and it is unlikely that one ever will be developed. For this reason a reasonable number of options have to be made available to handle the wide variety of problems facing design engineers.

Vapor–liquid equilibrium for most hydrocarbon systems can be adequately calculated by such generalized methods as Chao-Seader [89] or the many modifications of the Benedict-Webb-Rubin (BWR) [90] or Redlich-Kwong (RK) [91] equations of state. Enthalpy is well calculated using a version of corresponding states theory such as that developed by Pitzer [92] or Yen-Alexander [93] or by using an equation of state such as BWR.

Such generalized methods are typically applied in process calculations in an ethylene plant, although there are some situations where their application could lead to serious errors. For instance, the ethylene and propylene fractionators involve the separation of close boiling hydrocarbons. No generalized correlation will yield equilibrium data accurate enough for such systems, and specific data must be provided.

Specialized modular simulation programs are applied to determine the mass and energy balances for the refrigeration cascade and the steam system. Other computer programs have been developed to perform the process design calculations for various types of equipment including the pyrolysis heaters, fractionating trays, and heat exchangers.

7.4. Plant Safety

In all hydrocarbon processing plants, safety is of eminent importance. In olefins plants particularly, safety of design, both process and engineering, safety in operation, and safety in maintenance are of ultimate importance to commercial success.

Operating conditions in olefins plants reach extremes on both sides of the temperature and pressure scale, ranging from −160°C to over 1000°C and 3.45 and 12.4 MPa (on the steam side) respectively, requiring control instruments of many types and different ranges. From the time of concept of a project onward, the concern for safety is ever-present through the design stage, in operations, and in maintenance planning. It concerns the owner-operator and the engineer-contractor in equal measure, and it is not something for which provision can be made at a later date. Safety has to be built in from the ground up.

The subject has been extensively dealt with publicly by many authors. It has also been dealt with on a private basis in seminars sometimes organized by contractors for the benefit of information exchange between their principal clients and themselves.

7.4.1. Design

During the design stage, safety considerations pertaining to the selection of materials and instrumentation, safe spacing and accessibility, etc. are taken care of by the plant design team on the basis of existing codes and past experience. During this stage the design team has the assistance of metallurgical experts, instrumentation and environmental specialists, and safety engineers. Before the design is completed, a hazards and safety review is normally arranged at which all concerned, including the client's representatives, participate. During this review all phases of the plant's operations are consequentially analyzed, which eventually results in the preparation by the contractors operating department of detailed manuals for the start-up phase, normal plant operation, and shutdown phase. The reviewal team normally considers four basic contingencies that relate to safety and dependability at all plant operating levels: (1) containment and operating control, (2) safe release and disposal of liquids and vapors, (3) ignition prevention, and (4) incident control and disaster prevention. The last contingency relates to measures and procedures to prevent a minor incident in one area from escalating to neighboring areas, i.e., of localizing the potential damage to the plant.

Nonscheduled or crash shutdowns provide the ultimate test of the perceptive thinking and logical reasoning that is to be done by the hazards and safety review team.

Provision must be made to automatically isolate certain plant sections temporarily, such as the pyrolysis section, for example, in case of a local fire in or near the heaters, while the purification section may continue to operate under conditions of total reflux.

As another example, if a more serious emergency should command a crash shutdown of the plant, it will be imperative to shut down the feed and stop all heat supply while cooling water flow must continue to remove heat from the plant. Under such circumstances special dousing means are necessary to prevent damaging the heaters. Thus the objectives of the safety review toward the end of the design stage are numerous and complex [196].

7.4.2. Operations

The realization of plant safety in operations is largely dependent on the quality and competence of the operating and support personnel. The operating organization is normally made up of the manager of operations, qualified by years of experience in olefins plants, who is supported by a competent, more or less permanent staff of engineers that watches the performance of individual plant units from day to day. On the executive side he directs operations through an assistant and/or plant foreman. The latter has reporting to him the assistant foremen or supervisory shift operators in charge of the various plant sections, such as the heater, compression, purification, and utilities section. The total operating complement may reach 60 to 90 people (including staff) depending on plant size and complexity. All of these people must be fully conversant with plant layout and operations.

The problem of recruiting and training a skilled and competent plant operating staff thus is a formidable one. For the initial start-up a client may make arrangements to have a selected crew train at an existing plant, preferably one designed by the same engineer-contractor that designed and erected his own plant, and he may depend on the latter's operating staff to fill a number of key posts for the duration of an initial operating period. A client already in the business of operating a plant finds himself in a better position, as he can add trainees to his existing plant and withdraw at an early date at least a core of experienced operators for the start-up of a new plant. In either case, training and indoctrination must be intense, logical, and detailed to have full impact on the trainees.

The training curriculum normally starts with the study of flow sheets and operating manuals by the medium of a plant scale model. Some companies make it a practice to have their training crews participate in testing and preparing the plant for start-up by assisting in line and instrument testing, regenerating freshly loaded catalyst, and similar chores. Training programs, which may include brief references to the physical phenomena basic to an understanding of the various operating steps and of the logistics of the safety systems, are normally continued and intensified after start-up. Some operating companies develop programs during the early life of the plant of rotating operating personnel to instill in them an appreciation of the cohesion and interdependence of the sections of the entire plant complex.

In the final analysis, the success of start-up and normal or emergency shutdown maneuvers depends on the presence of mind, emotional stability, knowledge, and prior experience of the plant supervisory and operating personnel. The chance of human error, which no amount of automation can totally eliminate, remains ever-present.

7.4.3. Maintenance

Problems, many of them bearing on plant safety, arise in an operating plant that cannot be treated in a "static" manner. To mention a few, instrument leads may become inoperative, safety valves may not fully reset, and lines carrying polymerizable materials may plug up through inattention. Certain parts in the plant wear out faster than others and require more frequent attention. Statistically, the compressors are the most notorious cause of unscheduled plant shutdowns; their lube oil systems and monitoring instrumentation require constant supervision.

It is for the maintenance crew to develop programs of "preventive" maintenance to forestall the occurrence of incidents that might, if left unattended at the source, impact on the safety and dependability of sections or of the plant as a whole. The maintenance crew, at times assisted by staff engineers, is charged with developing statistical records of the imminent and actual failure of plant components, be it of pump seal leaks or heater tube wear or drainage and blowdown equipment, which will determine the scheduled frequency of inspections to maintain the plant in safe and dependable condition.

It is beyond the scope of this paragraph on plant safety to provide technical details on the many measures and procedures to promote and enhance plant safety. An extensive literature exists for the benefit of the reader intent on pursuing this subject further [197–200].

8. The Pyrolysis of Hydrocarbons

The pyrolysis of hydrocarbons for the production of petrochemicals is almost exclusively carried out in tubular coils located in fired heaters. Steam is added to the feedstock to reduce the partial pressure of the hydrocarbons in the coil. The reactions that result in the transformation of saturated hydrocarbons to olefins are highly endothermic and require temperatures in the range of 750 to 900°C, depending on the feedstock and design of the pyrolysis coil.

Pyrolysis methods other than externally heated tubular coils such as fluid beds and recuperators have been used industrially but have not achieved prominence [94].

Despite many years of research in universities and industry, no unified and comprehensive theory of pyrolysis has as yet been developed nor can the development of such a theory be considered likely. A comprehensive mathematical model of pyrolysis would permit the optimum design of pyrolysis reactors and the accurate prediction of pyrolysis yields for any hydrocarbon feedstock. Yet the enormous multiplicity of chemical reactions coupled with nonuniform heat transfer would appear to make the development of a comprehensive pyrolysis theory a herculian task. Most concepts applied to the design of pyrolysis reactors and the prediction of pyrolysis yields are of an empirical nature.

Important from a practical point of view is an understanding of the olefins formation from typical hydrocarbon feedstocks and secondly concepts that can aid in the design of optimum pyrolysis reactors for any feedstock or sets of feedstocks. Variations in the composition of feedstocks have a much greater influence on the pyrolysis yield spectrum for any given reaction system than have variations in the design and operating conditions of pyrolysis reactors. Most industrial tubular coil-type pyrolysis reactors are of very similar design, and their operating conditions, such as total and hydrocarbon partial pressures, temperatures, heat flux, and mass velocities, are within relative narrow ranges which have been found empirically to be most advantageous.

8.1. Free Radical Mechanism

Rice and Herzfeld [95] proposed the concept of the free radical mechanism as a basis for the decomposition of hydrocarbons, and this concept is generally accepted as correctly representing the sequence of pyrolysis reactions. It postulates three basic types of reactions: (1) chain initiations or the initial formation of radicals, (2) chain propagation through reaction of radicals with molecules, and (3) chain termination causing the disappearance of radicals. For paraffins the free radical mechanism may be described as follows:

Chain Initiations:

$$C_nH_{2n+2} \longrightarrow C_mH_{2m+1}^{\cdot} + C_{(n-m)}H_{2(n-m)+1}^{\cdot} \qquad \text{with} \qquad n \geqslant 2,\ m \geqslant 0$$

Chain Propagations:

$$C_nH_{2n+2} + C_mH_{2m+1}^{\cdot} \longrightarrow C_nH_{2n+1}^{\cdot} + C_mH_{2m+2} \qquad \text{with} \qquad n \geqslant 2,\ m \geqslant 0$$

$$C_nH_{2n+1}^{\cdot} \longrightarrow C_mH_{2m} + C_{(n-m)}H_{2(n-m)+1}^{\cdot} \qquad n \geqslant 2,\ m \geqslant 0$$

Chain Terminations:

$$C_nH_{2n+1}^{\cdot} + C_mH_{2m+1}^{\cdot} \longrightarrow C_nH_{2n} + C_mH_{2m+2} \qquad \text{with} \qquad n \geqslant 0,\ m \geqslant 0$$

$$C_nH_{2n+1}^{\cdot} + C_mH_{2m+1}^{\cdot} \longrightarrow C_nH_{2n+2} + C_mH_{2m} \qquad n \geqslant 0,\ m \geqslant 0$$

$$C_nH_{2n+1}^{\cdot} + C_mH_{2m+1}^{\cdot} \longrightarrow C_{n+m}H_{2(n+m)+2} \qquad n \geqslant 0,\ m \geqslant 0$$

Chain initiations involve the cleavage of C—C bonds and the formation of radicals. Chain propagations proceed by reaction of a radical with a molecule resulting in the formation of the conjugated radical and molecule or by decomposition of a radical into an olefin and a smaller radical. Chain terminations result from the reaction of two radicals forming either a saturated molecule or an olefin and a saturated molecule. It has been postulated that the reverse reactions also occur.

The first chain propagation reaction involves the extraction of hydrogen atoms from paraffins. Rice proposed that the reaction rate of hydrogen atoms depends on their position in the molecular structure of the paraffin and increases with the order of position from the methyl groups [96].

Zdonik et al. [97] have presented a very detailed treatise on the role of the free radical mechanism in pyrolysis including reaction rates for a variety of hydrocarbons and primary, secondary, and tertiary hydrogen.

The free radical mechanism describes the primary decomposition of saturated hydrocarbons to olefins, methane, and hydrogen, and predicts the product distribution of single hydrocarbons, in particular, at lower feedstock conversions with reasonable accuracy. At higher conversions secondary reactions involving the decomposition of primary reaction products become of significance and these can to some extent also be described by the free radical mechanism. This theory, however, does not adequately explain the formation of mono- and polycyclic aromatics.

The multitude of reactions that occurs during pyrolysis of naphthas and gas oils with their componential composition largely unknown precludes application of the free radical mechanism for the prediction of pyrolysis yields. Instead, typical pyrolysis yields for broad classes of feedstocks are generally determined by experiments in laboratory, pilot plants, or are obtained from industrial plants and are correlated based on a variety of empirical parameters.

8.2. Single Component Feedstocks

8.2.1. Ethane

Ethane, which is one of the dominant feedstocks for ethylene manufacture, lends itself very well to analysis by the free radical mechanism. The ethylene yield obtained in industrial pyrolysis furnaces is in the range of 85 to 90% of the theoretically possible yield of the simple dehydrogenation reaction:

$$C_2H_6 \longrightarrow C_2H_4 + H_2$$

This high yield indicates that the following propagation reactions dominate in accordance with the free radical mechanism.

$$\begin{aligned} C_2H_5\cdot &\longrightarrow C_2H_4 + H\cdot \\ C_2H_6 + H\cdot &\longrightarrow C_2H_5\cdot + H_2 \end{aligned}$$

Methane is formed in somewhat significant quantities by the following reactions:

$$C_2H_6 \longrightarrow CH_3\cdot + CH_3\cdot$$
$$C_2H_6 + CH_3\cdot \longrightarrow CH_4 + C_2H_5\cdot$$

Traces of diolefins, aromatics, heavier olefins, and paraffins are found in the reactor effluent.

For the industrial pyrolysis of ethane, the equilibrium approach concept has proven to be useful. Schutt [98] has evaluated some commercial data and the laboratory data reported by Kinney and Crowley [99], among others, and found that the methane yield increases and the ethylene yield decreases with the approach to the equilibrium of the ethane dehydrogenation reaction. The equilibrium approach ε can be defined as:

$$\varepsilon = \frac{P}{K_p} \frac{(y_{H_2})(y_{C_2H_4})}{(y_{C_2H_6})}$$

where P = hydrocarbon partial pressure, Pa
K_p = equilibrium constant for the ethane dehydrogenation reaction at the coil outlet temperature, Pa
y = mole fraction of constituent in the hydrocarbon mixture of the coil effluent

Based on Schutt's data and the assumption that every decomposing mole of ethane not converted to ethylene will yield 2 mol of other pyrolysis products, the molal yield of ethylene per converted mole of ethane can be represented by the following formula:

$$Y_{C_2H_4} = \frac{1.67K_p}{P} \frac{1-\alpha^2}{\alpha^2} \left(\sqrt{1 + 1.2\frac{P}{K_p} \frac{\alpha^2}{1-\alpha^2}} - 1 \right)$$

where α = fractional disappearance of ethane
$Y_{C_2H_4}$ = yield of ethylene, moles per mole of ethane converted

Up to ethane conversions of approximately 70%, which is beyond the range of industrial significance, the formula accurately represents the data reported within a few percent. The formula also shows that higher equilibrium constants and lower partial pressures favor ethylene formation. Higher equilibrium constants require higher reactor temperatures which in turn result in higher reaction velocities and therefore call for shorter residence times to maintain a given conversion and thus obtain a higher ethylene yield. It was the work of Schutt that prompted designers of pyrolysis furnaces to move in the direction of short reaction times and low hydrocarbon pressures for optimum production of ethylene from any feedstock.

Although Schutt's equilibrium approach [98] and the simplified first-order kinetics [133–138, Section 8.4] for ethane pyrolysis have for a time been accepted in industry and have been proven to be very useful, they depart from the fundamental free radical mechanism and hence their usefulness is limited.

Recent theoretical models employing the free radical mechanism for ethane pyrolysis, reported by Dunkleman and Albright [155] and Sundaram and Froment [205], seem to do an excellent job of predicting the results over a wide range of conditions.

8.2.2. Propane

Volkan and April [100] have reviewed the pyrolysis of propane. They concluded that (1) the pyrolysis of propane has an overall reaction rate of the order of 1.0 to 1.5 and the Arrhenius frequency factor constants fall between 10^{11} and 10^{14} s^{-1} while activation energies vary from 51,000 to 68,000 cal/mol, and (2) propylene and nitric oxide inhibit propane pyrolysis.

Further features regarding the review of propane pyrolysis by Volkan and April are reported by Albright [206]. Important details on (1) the formation of coke (or carbon), carbon oxides, and heavy hydrocarbons; (2) surface effects, and (3) kinetics and modeling of propane pyrolysis are consolidated and discussed.

Mechanistic modeling has been attempted to describe the pyrolysis of propane. Edelson and Allara [101], Dunkleman and Albright [156], and Sundaram and Froment [205] have reported the development of kinetic models to predict feedstock conversion and product distribution for propane. Edelson and Allara's model was reported to consist of 240 reactions involving 19 different radicals and 20 nonradical species, but the details of the specific reactions were not given. Rate constants for these reactions used in the kinetic models were derived from published experimental and thermophysical data.

8.2.3. Butanes

Similar studies and experimentation have also been reported for the pyrolysis of *n*- and isobutane [100, 102–106], pentane [107, 108], hexane [108–110] heptane [108–111], octane [112], and nonane [113–115]. These latter paraffins, for butane and heavier, are of less practical interest as they are normally not available as feedstocks in their pure form.

Thus, while it is recognized that the first-order pyrolytic reactions of paraffins and in general of other types of hydrocarbons may be described as simultaneous and consecutive reactions of free radicals combined with those of a molecular nature, the complexity of this representation increases rapidly with the carbon number of the feedstock to be pyrolyzed, and the practical impetus for further development in this direction is lacking.

8.3. Naphthas and Gas Oils

Straight run naphthas and gas oils consist of a mixture of differently structured hydrocarbons, paraffinic, naphthenic, and aromatic, that have boiling points within a predetermined temperature range. One of the first questions that arises is: Can the behavior on pyrolysis of these hydrocarbon mixtures be accurately enough described by the use of pseudo-values such as average molecular weight, average boiling point and/or refractive index, or any combination of physical properties characteristic of their composition?

Hirato et al. [116, 117] have represented these hydrocarbon mixtures by a single pseudo-hydrocarbon with a molecular weight identical to the average molecular weight of the feedstock. Kinetic models for the pyrolysis of these hydrocarbon mixtures have been developed based on the formation of C_4 and lighter hydrocarbons. These models are empirical and highly dependent on the experimental data used to develop the models. Illes et al. [118] have developed a kinetic–mathematical model for naphtha pyrolysis in tubular reactors based on the combination of kinetic relationships describing the overall decomposition rate and equations representing the curve-fitted pyrolysis yield data.

The development of models involves a considerable expenditure of technical talent and computer expertise. Furthermore, the results obtained from models have to be calibrated, at least in a semiempirical manner, with experimental pyrolysis data. For design purposes, greatly simplified yield correlations are employed to determine the yield and product distribution from the pyrolysis of a specified feedstock. The yield correlations are generally only valid for a specific pyrolysis furnace design.

Pyrolysis yields vary widely according to the type and properties of the feedstock and of reactor operating conditions. The effect of the latter is described in Section 8.6.

The effect of feedstock composition on pyrolysis yields is illustrated in Tables 10 and 11 of Chapter 6 which show typical once-through pyrolysis yields at high severities for the feedstocks listed in Table 12. The analyses and properties of the feedstocks are characterized by their specific gravity, distillation data, and breakdown into paraffins, naphthenes, and aromatics, commonly referred to as the PONA analysis. It may also be helpful to know the origin of feedstock or the crude source.

The typical yields correspond to a given tubular reactor design operated at specific conditions. Variations in reactor configuration and in operating conditions will result in changes in the pyrolysis yields given. Additional data on pyrolysis yields have been reported in the literature [119–128].

Both naphthas and gas oils are principally composed of three fundamental types of hydrocarbon structures [129]:

1. Paraffinic chains, normal or branched, that are relatively easy to pyrolyze by C—C cleavage and subsequent free radical reactions. Paraffinic structures are characterized by a high hydrogen to carbon ratio, and when decomposed produce high yields of gaseous products and low yields of heavy aromatic oils.
2. Cycloparaffins, with and without side chains, that decompose by ring rupture accompanied by the formation of alkyl radicals. These compounds require higher activation energy levels than do paraffinic chains. Their hydrogen to carbon ratios are intermediate to paraffins and aromatics, as are the yields of light gases and heavy liquid products.
3. Aromatic ring structures that are highly heat stable. Benzene is considered

one of the most refractory compounds. The hydrogen to carbon ratios of these compounds are low and their yield of desirable olefins is negligible. Aromatic rings have long been known to be tar precursors through condensation reactions which cause deposition of coke in the cracking coil.

The effect of naphtha composition on pyrolysis yields has been correlated as a function of their concentrations of paraffins, naphthenes, and aromatics. These correlations are supplemented with information pertaining to the gravity and distillation data of the feedstocks. Additional information on the split between normal and isoparaffins and the concentration of cyclopentane homologs present in the naphthenes is often useful.

Due to the presence of polycyclic naphthenes and aromatics, gas oils are still more complex in their constitution than naphthas and the correlation of pyrolysis yield data is more difficult. Although a hydrocarbon-type analysis of gas oils would provide ideal information for their characterization, this information is rarely available. Thus the effect of gas oil composition on pyrolysis yields has been correlated with specific gravity, distillation data, and other physical indices.

To obtain detailed information on the chemical structure of a gas oil, Lohr and Dittmann [170] have developed a procedure which yields statistical information on the types of bonding of the components of a gas oil. The structural elements accounted for are (1) the number of aromatic rings, (2) the number of naphthenic rings, (3) the number of carbon atoms in paraffinic chains, (4) the number of methyl groups linked to paraffin chains, etc. These structural elements are determined by high-pressure adsorption chromatography and hydrogen nuclear magnetic resonance (HNMR) techniques to calculate the composite ethylene yield from the chemical structure of the gas oil components. While the procedure of Lohr and Dittmann thus shows some similarity with the BMCI correlation discussed below, it places considerably more emphasis on the type of component branching of a specific gas oil feedstock.

Green et al. [129] have reported that the U.S. Bureau of Mines Correlation Index (see Section 6.3) may be successfully used to correlate the quality of gas oil feedstocks and its effect on pyrolysis yields.

For straight chain paraffins the BMCI is uniformly near zero up to C_{15}, branched paraffins range up to 15 (increasing with the extent of branching), monocyclics range from 37 to 50 (depending on the number and length of side chains), and polycyclic aromatics are well over 100. The maximum practical ethylene yield obtainable from gas oils appears to be a relatively simple function of the BMCI [130]. The BMCI is generally not useful for the prediction of the ethylene yield of single component feedstocks. As an example, decalin (bicyclodecane) and *p*-xylene have BMCI's of 72.3 and 71.4, respectively. Yet *p*-xylene produces no olefins while decalin pyrolyzes with relative ease and forms significant amounts of light components, including ethylene.

Davis and Keister [131] have proposed corrections on ethylene yields as a function of the feedstock composition (Table 14).

The effect on the feedstock hydrogen content on ethylene yields is also illustrated in Fig. 31 [123]. These factors are applicable over a limited range of

TABLE 14

Feed Composition Variables	Sensitivity of C_2H_4 Yield, ΔY, to Change in Feed Composition, wt.% of Feed
Isoparaffins in feed (replacing *n*-paraffins)	$\frac{\Delta Y}{\Delta(\text{iso})} = 0.18$ wt.%/wt.% iso in feed
Cyclopentane homologs replacing cyclohexane homologs	$\frac{\Delta Y}{\Delta \text{CP}} = -0.1$ wt.%/wt.% cyclopentanes in feed
% H (or H/C)	$\frac{\Delta Y}{\Delta \text{H}} = +5$ wt.%/wt.% H

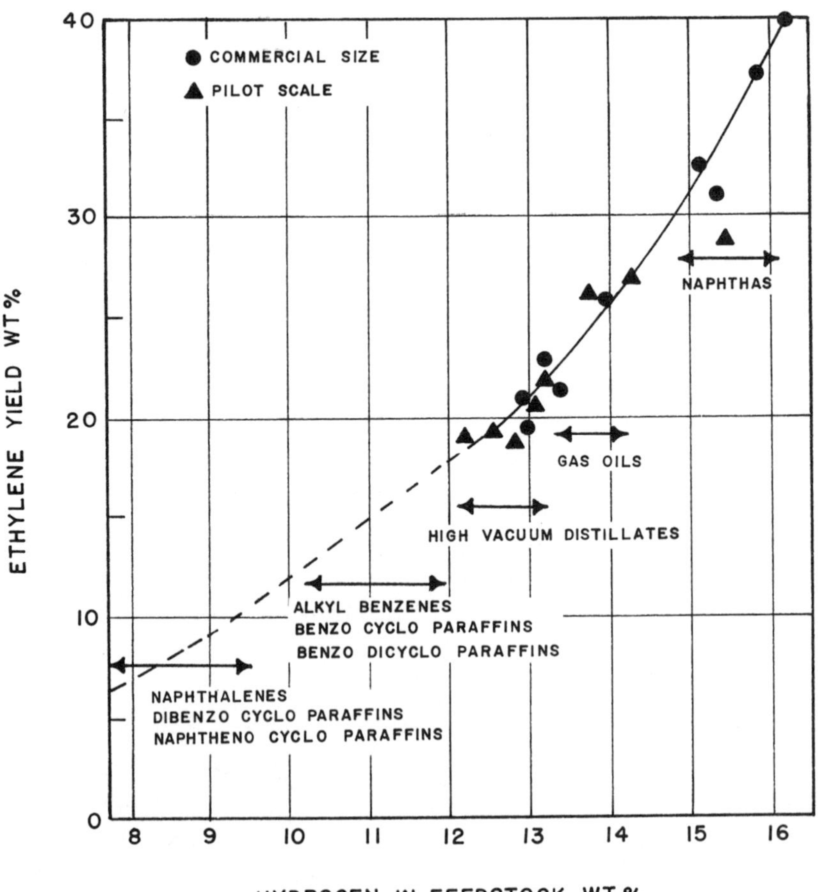

FIG. 31. Relation between hydrogen content of feed and single pass ethylene yield.

feed composition changes at identical operating conditions. The same authors also provided sensitivity factors for operational variables which are discussed below.

8.4. Kinetics

Most kinetic relationships developed to aid in the design of pyrolysis furnaces are of an empirical nature, based on pilot-plant work and data from industrial installations. Little consensus exists among various designers of pyrolysis furnaces as to which definitions of different parameters are most useful.

Residence Time. The so-called bulk residence time, which is most often mentioned in the literature to characterize various furnace designs, represents the entire length of time that a unit of gas spends in the pyrolysis coil. It is defined by

$$\theta_B = \int_0^L \frac{dl}{v(l)}$$

where L = total coil length, m
l = distance from the coil inlet, m
$v(l)$ = gas velocity, m/s
θ_B = bulk residence time, s

Postulating that the residence time of the pyrolysis products rather than the bulk residence time influences the yield spectrum, Fernandez-Baujin and Solomon [132] have proposed a so-called average residence time as a suitable parameter for correlation purposes. They define the average residence time by

$$\theta_A = \frac{1}{\alpha_0} \int_0^{\theta_B} \alpha(\theta)\, d\theta$$

where θ_A = average residence time, s
$\alpha(\theta)$ = fractional conversion of the feed
α_0 = fractional conversion at the coil outlet
θ = residence time, s
θ_B = bulk residence time, s

Zdonik et al. [133] also emphasize that the bulk residence time is of little significance without consideration of the time–temperature relationship and introduced the so-called equivalent residence time:

$$\theta_E = \frac{1}{K_5(\theta_B)} \int_0^{\theta_B} K_5(\theta)\, d\theta$$

where θ_E = equivalent residence time, s
θ_B = bulk residence time, s
K_5 = reaction rate of *n*-pentane, s^{-1}

The equivalent residence time represents the time in which n-pentane through an isothermal reaction at the coil outlet temperature would have been converted to the same degree as in an industrial pyrolysis coil where the gas temperature increases over a longer period of time.

Hydrocarbon Partial Pressure. The typical profile of the hydrocarbon partial pressure in pyrolysis coils is somewhat peculiar, in particular for the heavier feedstocks which require higher ratios of dilution steam. At the coil entry the hydrocarbon pressure is low and increases rapidly as the cracking reactions proceed, reaches a maximum, and then decreases toward the coil outlet. The total pressure, of course, decreases steadily from inlet to outlet.

Higher hydrocarbon pressures promote undesirable secondary reactions. Analagous to the average residence time and for the same reasons expounded by Fernandez-Baujin and Solomon, the average hydrocarbon partial pressure is defined as

$$\pi_A = \frac{1}{\alpha_0} \int_0^{\alpha_0} \pi(\alpha)\, d\alpha$$

where π_A = average hydrocarbon partial pressure, MPa
$\pi(\alpha)$ = hydrocarbon partial pressure as a function of conversion, MPa
α = fractional conversion

The hydrocarbon partial pressure at any given point in the coil is a function of the coil outlet pressure, the dilution steam ratio, and the gas velocity profile between the given point and the coil outlet.

Conversion. For individual components, conversion is simply defined as the fractional disappearance of the reactants. For complex multicomponent feedstocks of essentially unknown componential composition, this definition becomes more involved. For these reasons a conversion based on the molecular weights of pyrolysis feed and effluent is derived from the material balance:

$$\alpha = \frac{\dfrac{m_f}{m_e} - 1}{\gamma - 1}$$

where α = conversion
m_f = molecular weight of feed
m_e = molecular weight of effluent
γ = molal expansion factor (moles of effluent per mole of feed converted)

The molal expansion factor, although certainly dependent on the conversion, is usually assumed to be constant for any given feedstock.

Severity. The term severity, with adjectives mild, medium, and high, is often used descriptively to indicate how severely, i.e., mostly at what temperature, a feedstock is subjected to pyrolysis. Zdonik et al. [133] have quantitatively defined a so-called kinetic severity factor:

$$\text{KSF} = \int_0^{\theta_B} K_5 \, d\theta$$

where KSF = kinetic severity factor
K_5 = reaction rate of n-pentane, s^{-1}
θ = residence time from coil inlet, s

The "kinetic severity factor" and the "equivalent residence time" are closely related:

$$\theta_E K_5(\theta_B) = \text{KSF}$$

The logic behind using n-pentane as a reference hydrocarbon is that it is generally present in typical full range naphthas and that, because it is one of the most refractory components, its conversion can be measured in an industrial plant if sufficient analytical facilities are available. The usefulness of the kinetic severity factor is probably restricted to naphtha feedstocks.

Severity and conversion are in reality different terms for the same pyrolysis conditions; high conversion requires high severity, and high severity implies high conversion. For n-pentane the following relationship exists between the two parameters:

$$\alpha_5 = 1 - e^{-\text{KSF}}$$

Davis and Keister [131] and Leonard et al. [183] have reported on the use of other expressions such as the methane, ethylene, and propylene yields, the methane/propylene and ethylene/propylene ratios, and hydrogen content of the liquid products to represent cracking severity. The application of these severity indices depends greatly on the characteristics of the feedstocks, operating conditions of the reactor coil, and the reliability of the effluent analyses. Sometimes two or more severity indices are used to represent cracking severity.

The term "maximum severity" is often used but rarely defined. The term is only meaningful for a given tubular pyrolysis reactor and a given feedstock, and even then represents only an ill-defined limit on the severity of industrial pyrolysis such as to obtain maximum ethylene yield consistent with acceptable run length of the pyrolysis modules between shutdowns for the removal of coke deposits. Sometimes a minimum hydrogen content of the heavy pyrolysis products is used to define the maximum severity as the hydrogen content of these products appear to be suitable indicators of the coking propensity. Ennis, Boyd, and Orris [127] have called attention to the hydrogen to carbon atomic ratio of the liquid pyrolysis products as a valuable index of pyrolysis severity. Numbers of the hydrogen/carbon (H/C) atomic ratio in the pyrolysis gasoline product or C_5 and heavier products encountered in pyrolyzing various fractions of distillate feedstocks were given. Much higher severities are common in pyrolysis systems designed for the production of acetylene.

Selectivity. As in the case of conversion, selectivity is not readily definable for all but the lightest feedstocks. In the case of ethane pyrolysis, selectivity is simply defined as the molal or mass yield of ethylene per mole or unit mass of ethane converted. In the pyrolysis of most feedstocks, however, the yield of by-products is significant and the selectivity to ethylene alone cannot be the single governing parameter. In the pyrolysis of propane, for example, the yield of propylene is significant, and it would generally be of little industrial interest to increase the ethylene yield at the expense of a large drop in the propylene yield. Fernandez-Baujin and Solomon [132] have proposed the use of the yield ratio of methane to ethylene as a selectivity parameter. Schutt [98] has shown that the methane yield in ethane pyrolysis is a good indicator of selectivity to ethylene. Hydrogen balance considerations also support the usefulness of the methane yield as a suitable selectivity parameter since methane is the hydrocarbon with the highest hydrogen to carbon ratio and, therefore, high yields of methane from almost any feedstock imply high yields of products with lower hydrogen content than the light olefins.

For the purpose of pyrolysis furnace design, simple kinetic relationships are generally employed, although the possibility of using a more sophisticated model has been explored [207]. In practice, the pyrolysis reactions are considered to be first-order reactions. For a tubular pyrolysis reactor operating under plug-flow conditions, the interrelationship between the reaction velocity constant, residence time, and the feedstock conversion can be represented by

$$\int_0^{\theta_B} k\, d\theta = \ln\left(\frac{1}{1-\alpha}\right)$$

where θ_B = bulk residence time, s
k = reaction velocity constant for a first-order reaction, s^{-1}
α = conversion

The reaction velocity constant k is dependent on temperature in accordance with the Arrhenius equation:

$$k = Ae^{-E/RT}$$

where A = frequency factor
E = activation energy

Reaction velocity data for a variety of hydrocarbon systems ranging from ethane to gas oils have been published in the literature. Based on first-order kinetics, these data have been plotted in Fig. 32 for ethane [133–138], naphthas [118, 137, 139–141], and gas oils [116, 137, 141] covering a wide range of cracking severities.

Generalized correlations of first-order rate coefficients for paraffins and naphthenes have also been presented by Nelson [137], Zdonik et al. [133, 142], Murata et al. [143], Davis et al. [144], and Voge et al. [145].

The first-order kinetic model concerning the disappearance of hydrocarbon feedstocks previously discussed does not account for the secondary,

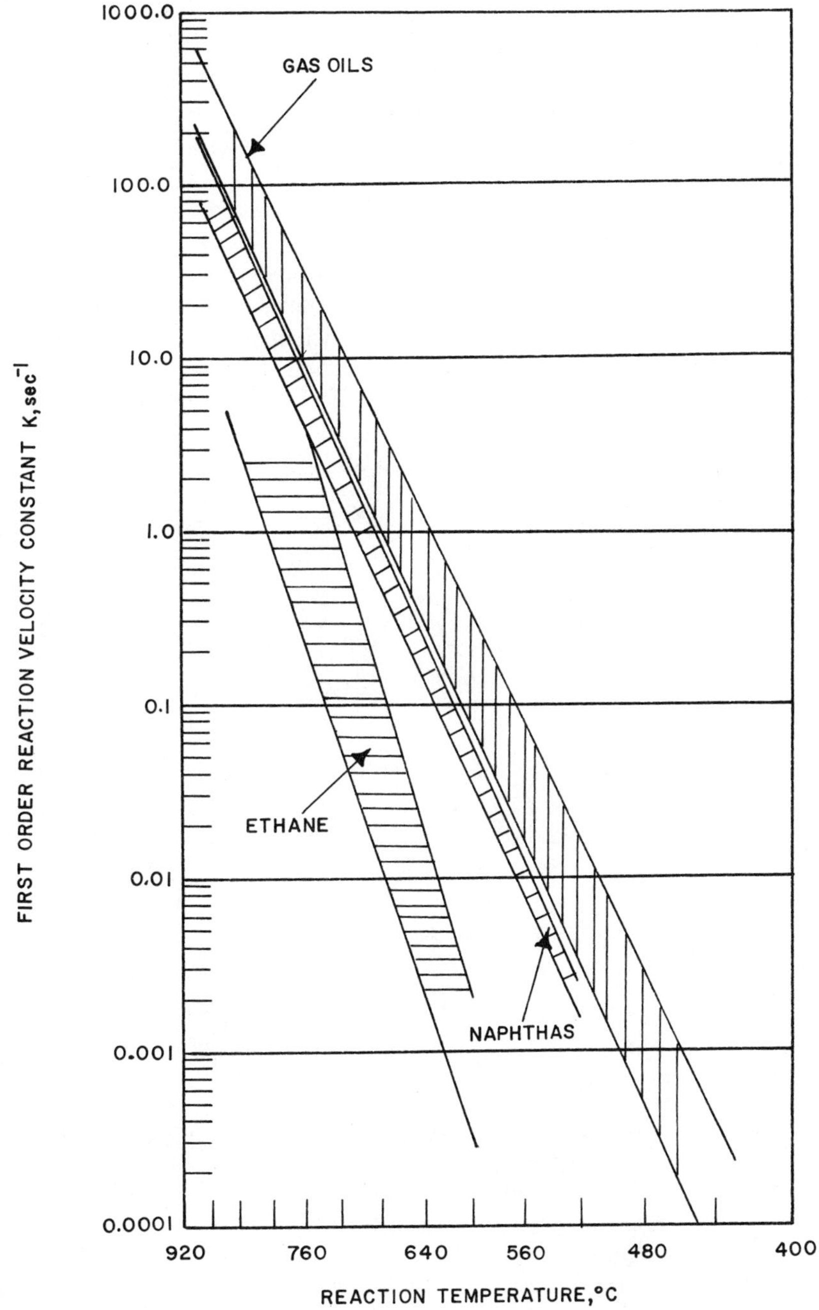

FIG. 32. First-order reaction velocity constants vs temperature for the pyrolysis of ethane, naphthas, and gas oils.

parallel, reverse, or other reactions occurring in the pyrolysis reactor. Several approaches to correct this deficiency have been investigated. Buekens and Froment [106, 146] and Murata et al. [143] have accounted for these effects by a hyperbolic relationship:

$$k = \frac{k^0}{1 + a\alpha}$$

where k = actual reaction velocity constant, s^{-1}
k^0 = apparent first-order reaction velocity constant, s^{-1}
a = inhibition coefficients which are dependent upon the hydrocarbon system and the reaction temperatures
α = conversion

Others have also accounted for these effects by using an empirical equation:

$$dc/d\theta = -k^* c^\eta$$

where c = concentration of the reactants
k^* = pseudo-reaction velocity constant
η = reaction order
θ = time

Both k^* and η are empirically determined from experimental data.

Some investigators correlate average molecular weight against experimental yield data to obtain the pseudo-reaction velocity constant k^*.

These empirical kinetic relationships permit a determination of feedstock conversion but will not provide adequate information on product distribution of any particular pyrolysis effluent.

8.4.1. Coke Formation and Deposition

Of particular interest are kinetic models that permit prediction of the coke deposition rate. The onstream factor of a pyrolysis reactor is determined by the fouling (coking) of either the pyrolysis coils or the effluent transferline exchangers. The operation of pyrolysis coils is normally terminated by the limiting metal wall temperature and/or an increase of coil inlet pressure. The former is related to the tube service life while the latter is related to the deterioration of the desired pyrolysis yields. The information on the variation of metal temperatures and inlet pressure with time on stream, both resulting from coke laydown, is used to determine the optimal combination of high cracking severity and coil throughput. To achieve this goal requires methods to predict the coking rate in pyrolysis coils.

Several studies have been reported on the formation mechanism of pyrolytic carbon. Slater [147] and Lobo et al. [148] suggested that the decomposition of hydrocarbons on a heated surface takes place as a direct decomposition of hydrocarbon molecules into carbon and hydrogen. Lanett et al. [149], Conroy et al. [150], Grisdale et al. [151], Appleby et al. [152], John et al. [153], and Syskov [154] have proposed mechanisms of coking. In general, most investigators agree that pyrolytic carbon or coke is mainly composed of high molecular hydrocarbons of an aromatic nature.

Dunkleman, Tsai, Brown, and Albright [155–158] have reported on surface reactions including the carbon deposition occurring during pyrolysis of light hydrocarbons (ethane, ethylene, propane, and propylene). These investigators developed coke yield correlations as a function of residence time and temperature. The results are in agreement with those reported by Kinney and Crowley [159] on the pyrolysis of ethane, ethylene, propane, and propylene in a Vycor reactor. In summary, the yields of deposited carbon (or coke) increase with increasing residence time and temperature. However, it should be noted that in the small diameter reactors with high surface to volume ratios used in these investigations, the carbon yield is expected to be much larger than in commercial reactors.

Additional evidence on coke formation has more recently been reported by Albright et al. [208–210]. Factors affecting coke formation include the material used for the reactor coil, past history and pretreatment of the reactor, and operating conditions. Coke is apparently formed by two mechanisms. The metal surface of a reactor catalyzes the growth of a filamentary-type coke containing metal granules and is magnetic. Second, coke is formed by condensation, polymerization, or agglomeration in the gaseous phase. This type of coke is nonmagnetic in nature.

Shah et al. [160] reported coking formation on the thermal cracking of *n*-octane. During the beginning of the run, the rate of coke formation is high. The coking rate continuously decreases with onstream time until a steady-state value is achieved. Apparently the high coking rate during the initial period of the run is due to the surface effects of the reactor tube walls. As the surface becomes covered with a coke layer, the surface effect diminishes. The coke formed during the later "steady-state" period is most probably thermal coke, i.e., formed as a result of homogeneous reaction rather than that formed on the surface. Figures 33 and 34 illustrate the initial and the steady-state coking rates reported by Shah et al. Similar behavior in coking rate has been observed in commercial size reactors.

Coking in a commercial pyrolysis coil is a highly complex process which has not been modeled in precise mathematical terms. However, simplified models utilizing certain empirically determined constants have been used to describe the coil coking phenomenon.

De Blieck and Goossens [161] developed a model which is based on the formation of coke in the laminar region near the tube wall of the pyrolysis coil. Mol [162] presented operating conditions and run length data for naphtha and gas oil cracking in commercial heaters. He concluded that the coking of pyrolysis coils is primarily due to the film effect for low mass velocities and small tube sizes, but is mainly due to high tube wall temperatures for high mass velocities. In general, commercial pyrolysis coils are operated at high mass velocities. Chambers and Potter [163] reported the coking rate as a function of

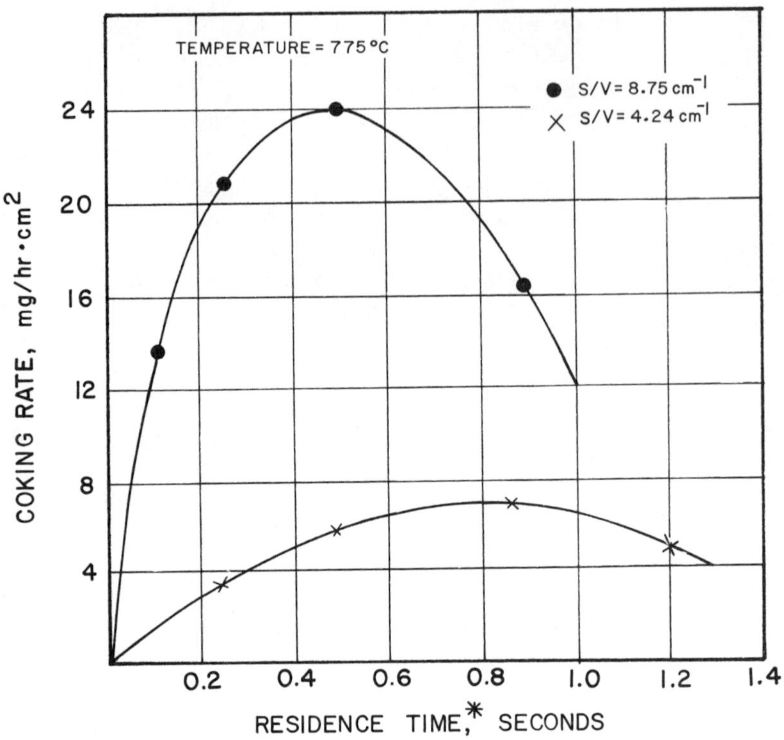

*In reference (160) called space time

FIG. 33. Initial coking rate vs residence time for thermal cracking of *n*-octane.

the pyrolysis fuel oil partial pressure, the cracking severity, and the mass velocity. The partial pressure of the fuel oil represents the concentration of coke precursors. The cracking severity and mass velocity define the temperatures in the process fluid stream, gas film, coke layer, and tube metal.

Fernandez-Baujin and Solomon [132] reported the development of a coking model based on the assumption of a two-step mechanism involving the mass transfer of coke precursors from the bulk of the gas to the walls of the tube, and the chemical reaction of coke precursors at the tube wall resulting in the formation and deposition of coke. Postulating that the reaction velocity constant is much higher than the mass transfer coefficient and therefore the rate of coil coking is mass transfer controlled, they developed a formula for the coke deposition rate as a function of mass velocity and pyrolysis tube diameter.

To reduce the coking problems in pyrolysis coils, it has at times been industrial practice to pretreat the reactor surface with sulfur compounds before the run is initiated and/or to add trace amounts of sulfur compounds during the normal course of the pyrolysis run to inhibit the catalytic activity of the tube material [162].

8.4.2. *Feedstock Diluents*

The effect of various pyrolysis feedstock diluents such as methane and particularly hydrogen on the kinetics of pyrolysis has at times received attention. Kunugi et al. [164] have carried out laboratory tests under atmospheric pressure using hydrogen and methane as diluents in various ratios to naphtha. Chahvekilian [165] has reported on laboratory work and the performance of a semicommercial prototype plant using a process called "hydropyrolysis." In this process naphtha and heavier feedstocks are pyrolyzed in the presence of hydrogen under pressures in the range of 2.0 to 3.5 MPa. The residence time is in the order of 0.075 to 0.1 s, and the maximum reactor temperature is close to 900°C. Hydrogen addition is in the order of 0.1 kg/kg of feedstock. The process is characterized by relatively high yields of methane, ethane, and propane, with the latter two to be recovered and pyrolyzed under standard low-pressure conditions. High yields of ethylene from any liquid feedstock and a reduction of the energy requirement for the pyrolysis gas compression are considered the most important advantages of this process.

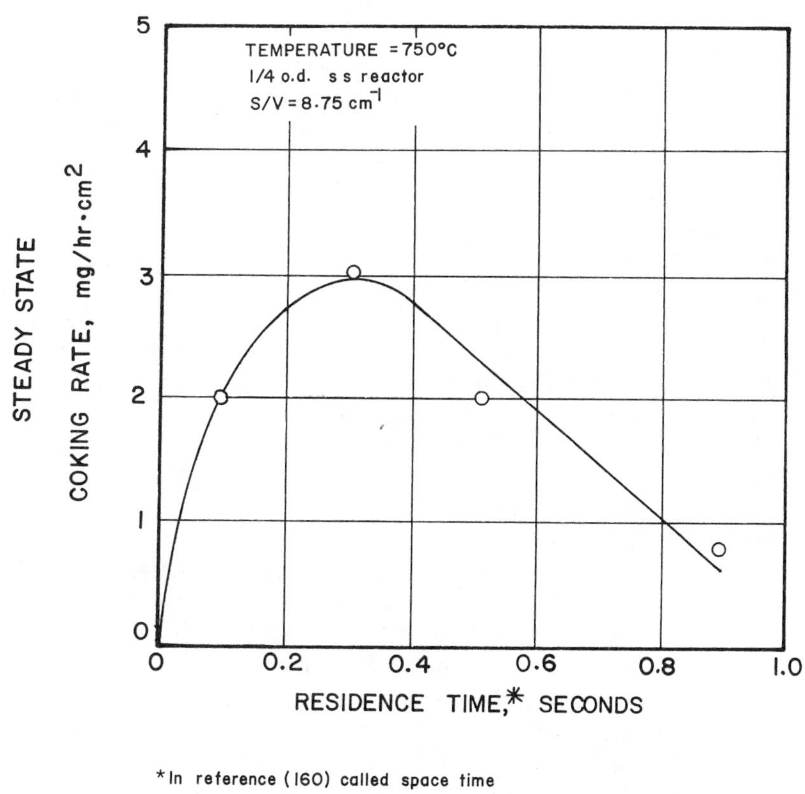

FIG. 34. Typical steady-state coking rate vs residence time for thermal cracking of *n*-octane.

8.5. Furnace Design Concepts

The typical modern pyrolysis furnace consists of a rectangular fire box with a single or sometimes a double row of vertical tubes located in the center plane between two radiating refractory walls. The heat transfer to the tubes is effected largely by radiation and only to a small degree by convection. The firebox temperature is typically in the range of 1200 to 1300°C.

It is the diameter and length of the tubes and the manner in which they are interconnected to form the pyrolysis coil which determine to what extent a particular design will be characterized by an optimum combination of pyrolysis parameters.

The design calculations applied to pyrolysis coils are necessarily complex since heat transfer and chemical reactions are involved. They are generally performed with the aid of computers and the use of iterative methods. Feed inlet temperature, i.e., the crossover temperature from the convection to the radiant section, and coil outlet pressure are usually fixed. The criteria applied to the selection of the crossover temperature have been discussed in Section 7.2.1 where a range of 550 to 700°C was also given. The coil outlet pressure is approximately 0.2 MPa or just high enough to avoid any danger of vacuum in the systems between the pyrolysis furnace and the main pyrolysis gas compressor.

The feedstock conversion at the coil outlet is an independent variable which the designer will select either directly, as in the case of feedstocks of known componential composition, or by calculation from the molecular weights of the feed and effluent. The effluent composition is determined as a function of feedstock characteristics on the basis of an empirical yield correlation valid for a particular class of furnace designs.

The heat of reaction can be determined either from the enthalpies of formation, if the componential composition of feed and effluent are fully known, or from the heats of combustion in the case of complex feedstocks. For the calculation of the major pyrolysis parameters as well as the determination of the profiles of conversion, hydrocarbon partial pressure, pyrolysis gas, and tube wall temperature, the pyrolysis coil is subdivided into sufficiently small sections for which equations describing the reaction mechanism, heat, mass, and momentum transfer can be written. Figure 35 shows the major profiles for a typical ethane pyrolysis coil.

Various options exist for the specific design of a pyrolysis coil, and this fact accounts for the variety of industrial pyrolysis furnaces presently in operation. If short residence time is considered the single most important objective, there will result a short coil with tubes of small diameter and in the extreme a single tube of 20 to 40 mm diameter and 12 to 15 m length. Such a coil will have a low capacity and a relatively short operating period between decokings because the performance of small tubes is more affected by coke deposits than larger tubes.

If a combination of high capacity, medium residence time, and low hydrocarbon partial pressure is judged to be most beneficial, a pyrolysis coil composed of relatively large tubes (100 to 200 mm) will result. Designs that aim for particular profiles of pyrolysis gas temperatures and hydrocarbon partial pressure will call for either a coil configuration composed of parallel tubes and

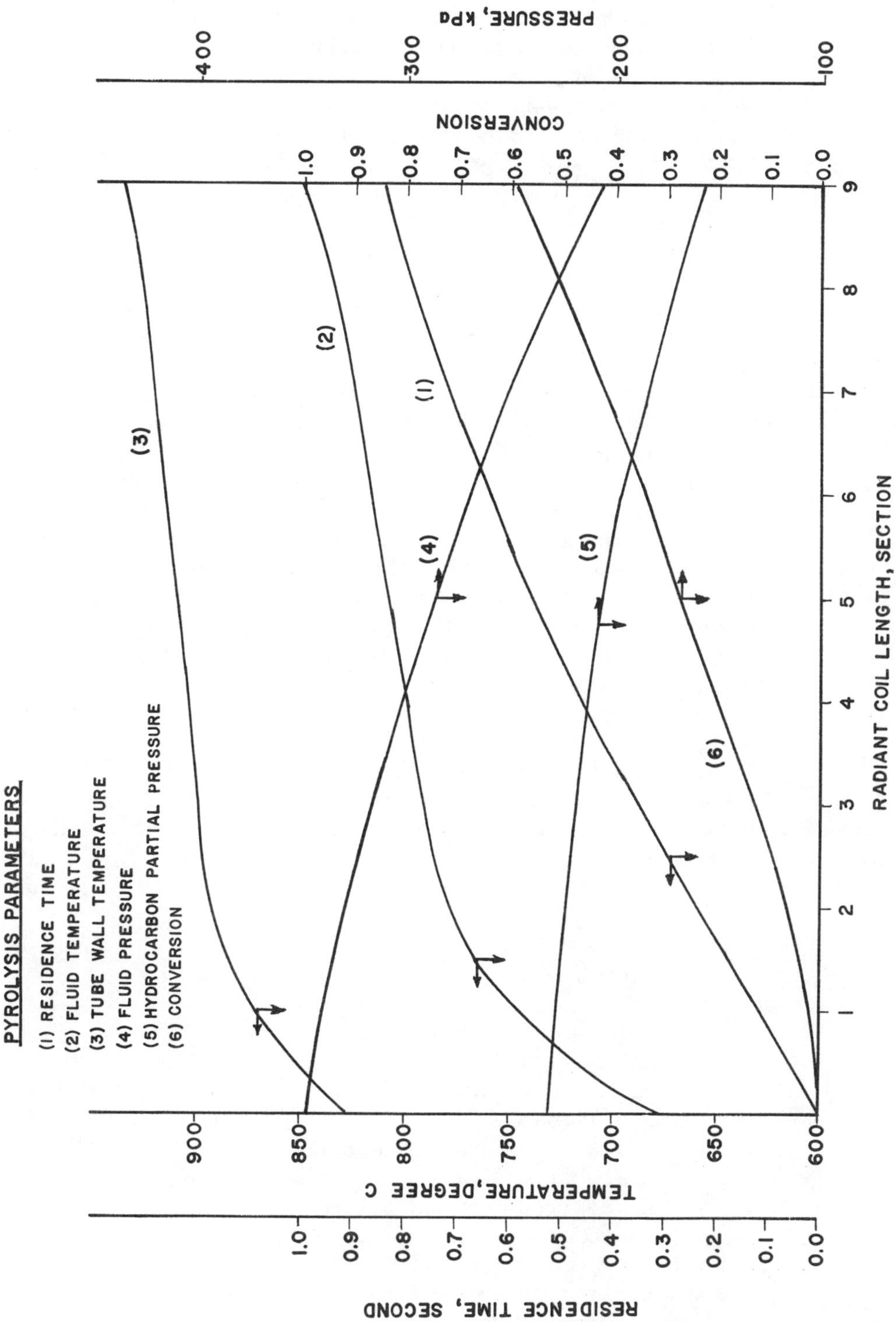

FIG. 35. Major profiles for a typical ethane pyrolysis coil.

tubes in series or separation of the firebox into zones that permit independent control of the heat supply to various sections of the coil.

If maximum furnace capacity and full utilization of the firebox is a prime objective, coils can be located in two rather than one plane in the center of the box. Due to the shading effect, this will result in somewhat less than ideal heat transfer to the tubes and therefore a lower capacity of the individual coils, but will nevertheless give a higher total capacity for the pyrolysis furnace.

8.6. Heaters and Transferline Exchangers

The pyrolysis heaters and their transferline exchangers constitute the most important parts of an olefins plant. Untold efforts over the years have been devoted to their efficient design and refinement by specialty firms such as Stone & Webster, Pullman-Kellogg, C-E Lummus, KTI, and Linde-Selas.

The design of pyrolysis furnaces requires the selection of nonoxidizing, carburization-resistant alloys and careful analysis of the coil suspension system as well as of the stresses due to thermal expansion and contraction.

The hydrocarbon effluent leaves the reactor coils at temperatures of 750 to 900°C. Tube metal temperatures are in the order of 100°C higher.

The high coil exit temperatures make it mandatory, for the sake of fuel efficiency, to recover the largest possible amount of sensible heat from the coil effluent. Furthermore, chilling or quenching of the effluent must take place rather rapidly to avoid the deteriorating effect on yields of additional residence time outside of the confines of the combustion chamber. In early plants the furnace effluent was quenched rapidly through injection of a circulating stream of fuel oil, a procedure that degraded the recoverable heat.

Most modern plants incorporate so-called transferline exchangers to cool the furnace effluent and to generate steam with a pressure of as low as 4.0 MPa in the case of light paraffin pyrolysis and as high as 12.5 MPa in the case of naphtha and gas oil pyrolysis. The higher steam pressure is required to minimize the condensation of tarry material in the transferline exchangers.

8.6.1. Typical Heater Configuration

The patent literature is replete with propositions on how to construct a pyrolysis heater that would best meet the exacting demands of a thermal reactor for the production of olefins. It is somewhat surprising to see that the exterior appearance of the successful designs that have evolved are nearly identical. Figure 36 shows the principal elements of this design with a top coil outlet, and Fig. 37 with a bottom coil outlet. The heaters are normally built as twin units or cells that share structural elements and a common stack. Each unit consists of an upper convection section and a lower radiant section. The radiant box of the unit heater may contain four and more radiant coils arranged lengthwise, any two of which are normally connected to one transferline exchanger.

The convection section contains at least two and may contain as many as five horizontal tube elements (as shown) running the full length of the heater to

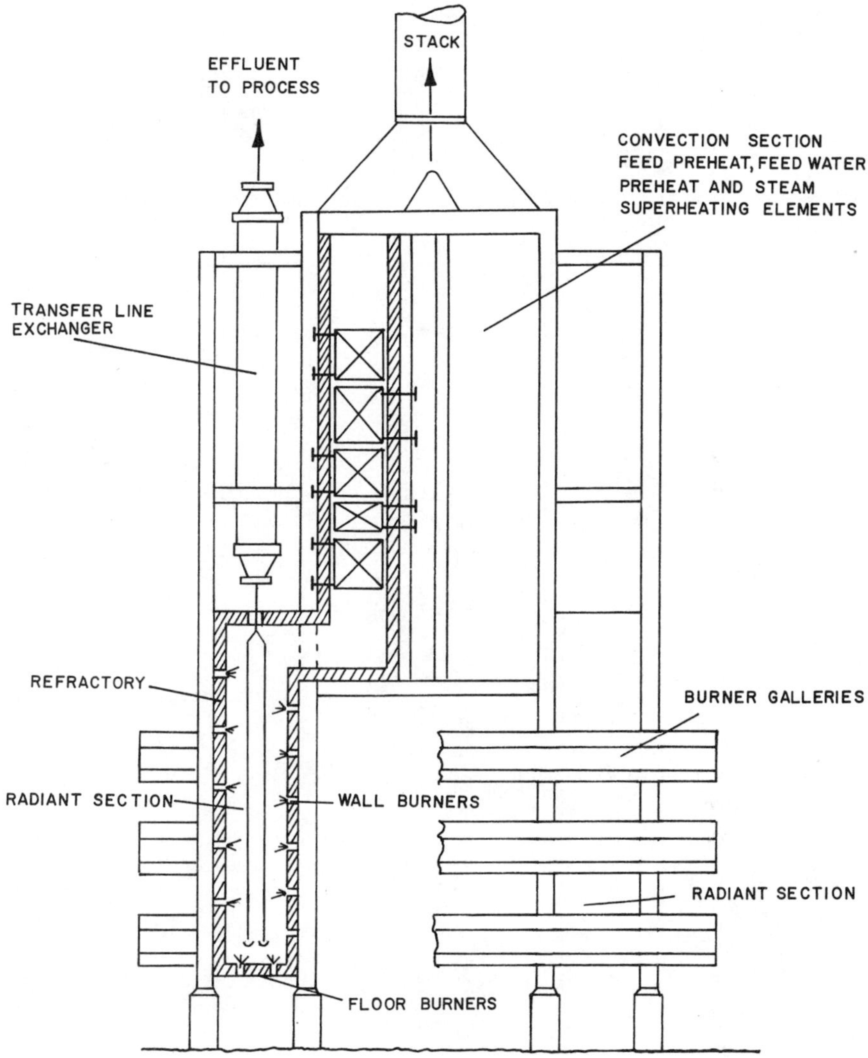

FIG. 36. Typical heater configuration—top coil outlet.

preheat the hydrocarbon feed, superheat dilution steam and high-pressure steam from the transferline exchangers, and preheat boiler feedwater to insure optimal utilization of the flue gas waste heat. Division of the total hydrocarbon charge over the number of heater coils generally takes place ahead of the heater, so the feed preheat elements will consist of as many parallel strings as there are reactor coils.

The reactor coils consist of a series of up and down lengths of tubing connected by return bends, located in the center plane of the fire box, and most often suspended on spring hangers from above the top of the radiant combustion chamber. The sections or passes of coil may be of uniform diameter

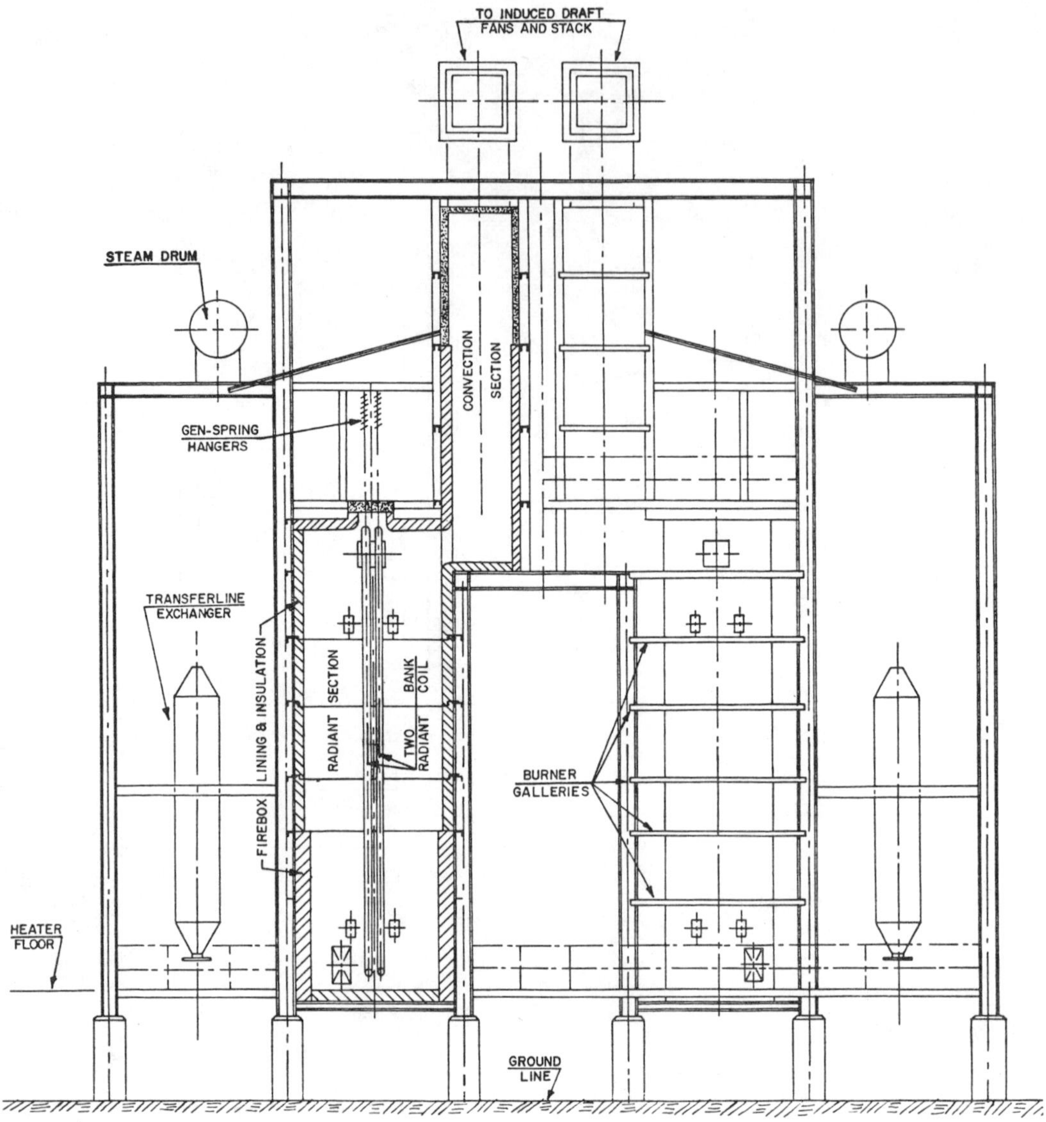

FIG. 37. Typical heater configuration—bottom coil outlet.

or may be swaged up to a gradually larger diameter toward the outlet end of the heater, or may be combined, as for instance, from several parallel passes to one pass (Fig. 38) when approaching the outlet, depending on what temperature profile is considered most desirable by the designer. The cross-sectional area determines the velocity, which in turn determines the actual feedstock residence time in the reactor coil.

Details on coil arrangement are generally considered proprietary information by the companies engaged in their conceptual design, and little factual information is available.

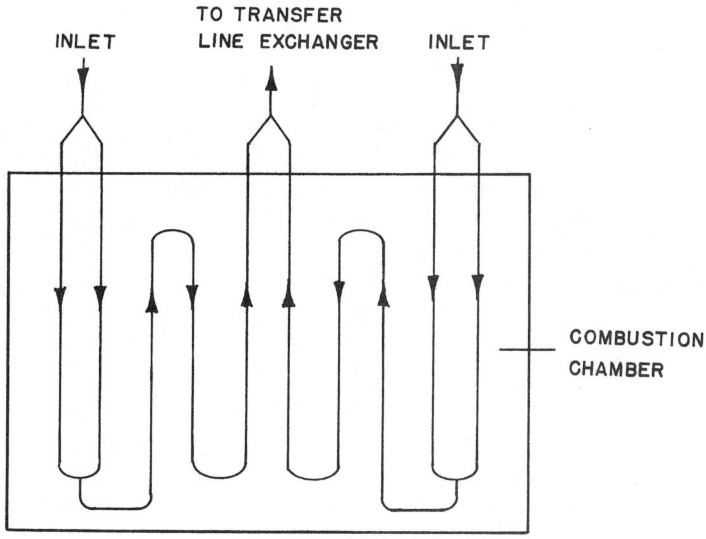

FIG. 38. Specimen reactor coil—short residence time heater.

For many years a special 25% chrome–20% nickel alloy with the designation HK-40 has been the preferred material for the tubes of the pyrolysis coil. The tubes are manufactured by a centrifugal casting process and the inside surface is generally machined to a specified smoothness. Machining of the tube results in a lower carburization rate. A maximum tube wall temperature of 1040 to 1050°C is typically used as the basis for design, and this results in an average life of the hottest tubes of 4 years or more. Tube failures are invariably a consequence of carburization. Clark [166] has reported carburization rates for HK-40 based on field data from several plants.

More recently, designers have turned to alloys with a higher nickel content and specifically HP-40 containing 25% chrome and 35% nickel. This material has a maximum service temperature of 1100°C and therefore permits coil designs for shorter residence times or higher capacity or a combination of both.

Table 15 presents a list of the major characteristics of pyrolysis furnaces; it covers most designs presently offered by various companies.

TABLE 15 Pyrolysis Furnace Characteristics

Ethylene production capacity, t/yr	20,000–70,000
Number of coils	2–20
Tube diameter, mm	50–200
Coil length, m	50–80
Gas outlet temperature, °C	750–900
Tube wall temperature, clean, °C	950–1050
Maximum tube wall temperature, °C	1040–1100
Average heat flux, kw/m^2	50–80
Residence time, s	0.15–0.6

The noticeably wide range of tube diameters is the result of different approaches taken by various designers. Extremely short residence times require short coil lengths and only small diameters provide the required surface to volume ratios. Alternately, larger tube diameters will result in longer residence times but also in lower average hydrocarbon partial pressures with approximately the same selectivity. The design of pyrolysis heaters is a complex task requiring careful balancing of residence time, pressure and temperature profiles, and rate of coke deposition, all within certain constraints imposed by the materials of construction. As a result, several different coil designs have evolved, and the industrially more important are to be described in further detail.

The combustion chamber housing the reactor coils is equipped with a multitude (in many instances well over 100) of air inspirating burners designed to produce a flat flame pattern to avoid flame impingement on the tubes and simulating a uniformly radiating wall. Induced draft is often used in instances where resistances on the flue gas side of the convection section make it necessary.

Heaters handling light hydrocarbon charge stocks are normally natural gas fired. Those in heavier feedstock service, such as naphthas and gas oils, may fire up to 30% of the fuel requirement as fuel oil, the balance as natural gas. It is claimed (Kellogg) that with one type of heater, totally floor-fired, up to 80% of oil fuel firing can be obtained. Economic pressures presently tend to force an increase toward a capability to use proportionately more of the cheaper oil fuel, and development work is underway to meet this trend. Fuel oils used must be low in ash (<100 ppm), sulfur (<1.0%), and particularly in vanadium (<5 ppm).

Lohr and Dittmann [186] have also pointed out that the viscosity of the fuel oil fired should be held constant, that it should not contain any particulate carbon, and that due to its instability it should not be held in storage for any length of time but burned as produced. With a top coil outlet, the vertical tube transferline exchangers are mounted as close as possible to the coil outlet. Horizontal elements set below the hearth have been used with bottom coil outlets, but the former type is much preferred.

8.6.2. Industrial Pyrolysis Furnaces

Stone & Webster is known to favor, for the "Ultra Selective Cracking" furnace, the use of small diameter tubes of 50 to 63 mm i.d. by ~10.0 m long (coil length = ~20 m) in U-tube vertical banks to achieve the short residence times that are conducive to high ethylene yields and selectivity when treating heavier liquid stocks. Each single pass or loop communicates directly with a transferline exchanger (USX) element of proprietary design, set above the roof of the combustion chamber and of the same or nearly the same tube diameter as the reactor exit tube. The temperature of the coil effluents is reduced by these elements to about 600 to 650°C. To achieve high heater capacities, a great number of individual reactor coils must be installed. A crossover distribution problem to these individual loops from the convection section is avoided through use of restriction orifices, and the direct coupling of the pyrolysis coil to its transferline exchanger minimizes the residence time at the highest gas temperature. For details concerning the logic of choice of small tube diameters, the reader is referred to a series of articles by Zdonik et al. [167].

The Pullman-Kellogg heater is the only one in the industry that offers a "multizone" design in which the radiant chamber is subdivided into three compartments, shielded from one another by refractory, through which the reactants and flue gases pass in parallel. The radiant chamber extends horizontally and the coils, as in other designs, are suspended vertically. This design permits the adjustment of the tube wall and gas temperature profiles through different firing rates in the three zones. Kellogg [168] has reported that so-called "concave" gas temperature profiles result in a significant ethylene yield enhancement. Taking stock of the generally recognized fact that short-time residence is beneficial, particularly with naphtha and gas oil feedstocks, Kellogg also developed a heater, offered to the industry under the name "Millisecond Heater," of which a prototype was installed in cooperation with Idemitsu Petrochemicals Co. in Japan in 1972. The millisecond heater design provides for a residence time of less than 100 ms (0.1 s) and reportedly lifts the ethylene yield on a naphtha, not further specified, from 29.3 wt.% for a conventional heater of Kellogg design to 33.3 wt.% with an attendant increase in outlet temperature of about 50°C [216].

C-E Lummus offers three heater models. As with most firms engaged in this field, the consecutive development of these models, known as SRT-I through SRT-III, was based on pilot-plant studies (1966 to 1972) that confirmed the short residence time phenomenon. The SRT-I model contained coils of uniform diameter. The later SRT-II model uses coils of increasing diameter toward the coil outlet or split coils that are joined toward the outlet. The SRT-III model represents a higher capacity version of the SRT-II. Both of the latter models are designed to operate with a "convex" temperature–length profile with but a slight inflection point near the coil outlet. It is claimed that the use of larger tube diameters produces operational and cost benefits from lower hydrocarbon partial pressures in a critical region of the coils as well as from a lower sensitivity toward carbon deposits and reduced residence time.

The capacities and residence times listed in Table 16 have been announced by Lummus.

TABLE 16 Lummus Heater Data

Model[a]	Residence Time (s)	Ethylene Capacity (t/yr)[b]
SRT-I	0.6–0.7	To 30,000
SRT-II HS (high severity)	0.3–0.35	To 25,000
SRT-II HC (high capacity)	0.5–0.6	To 45,000
SRT-III	0.27–0.45	50,000

[a]One cell = one radiant box plus convection section.
[b]Of ethylene from full range naphthas and atmospheric gas oils.

TABLE 17 Linde-Selas Heater Data

Model[a]	Tube i.d. (mm)	Reactor Coil Length (m)	Residence Time (s)	Ethylene Capacity (t/yr)
LSUC (uniform coil diameter)	50–125	60–70	0.3–0.4	50,000
LSSC (variable coil diameter)	Up to 150	70–80	0.3–0.4	70,000
LSCC (variable coil diameter)	60–150	45–60	0.2–0.4	70,000

[a]One cell = one radiant box plus convection section.

For a description and evaluation of Lummus pyrolysis heaters, the reader is referred to Refs. 169 and 171.

Linde A.G. and Selas Corp. of America have joined to offer large capacity heaters, suitable for ethane through vacuum gas oil feedstocks, embodying radiant coil designs oriented in one or two vertical banks. At the heater outlet four process passes are combined and led to one transferline exchanger. The radiant chamber may either be fully gas-fired through sidewall mounted burners or partially side-wall and floor-fired if liquid fuels are used in addition. The main parameters, other than feedstock specifications, taken account of in the coil design are partial pressure, residence time, severity, and film effects.

Table 17 lists characteristic dimensions and capacities.

The temperature-reactor length profile for model LSCC in comparison with a uniform diameter coil is shown in Fig. 39 [172].

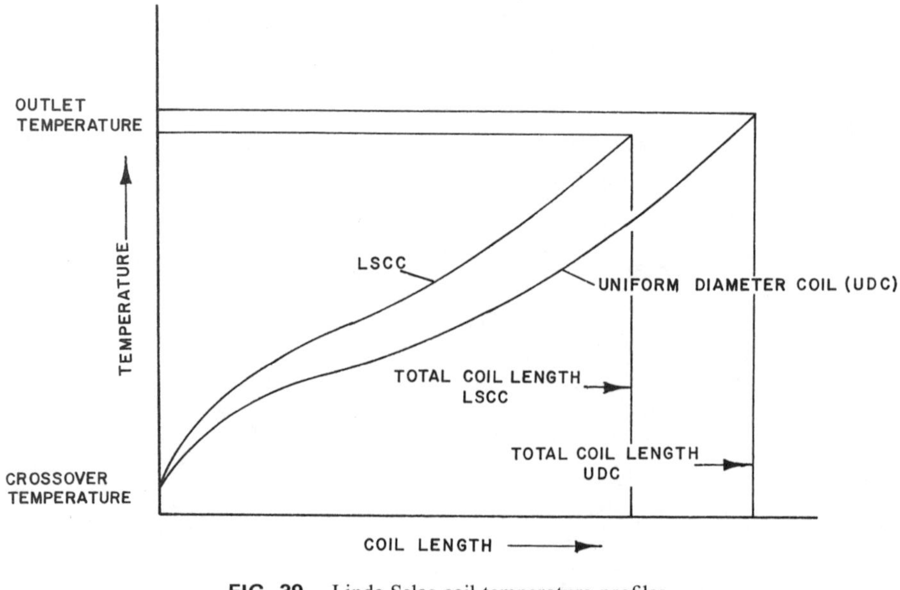

FIG. 39. Linde-Selas coil temperature profiles.

In Japan, Mitsubishi Petrochemical Co. and Mitsubishi Heavy Industries jointly developed a heater (Mitsubishi thermal cracking furnace = M-TCF) that in many respects differs from the heaters thus far described. Figure 40 presents a sectional view of this heater and Fig. 41 of the transferline exchanger used in conjunction with it.

In this design the convection section is arranged below the radiant box, and flue gases leave by way of a flue gas duct below the heater structure. The heater

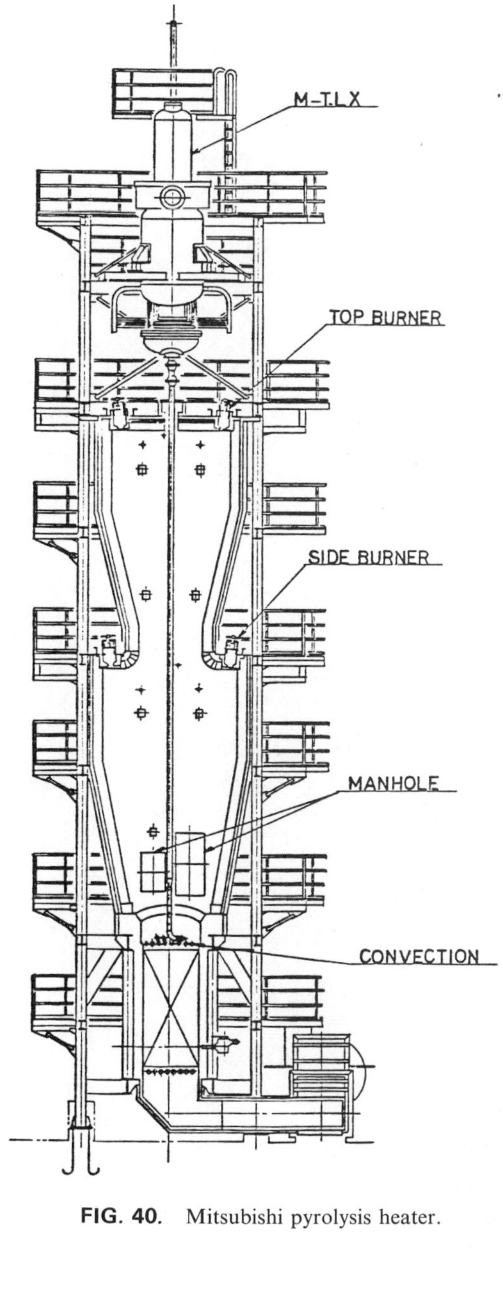

FIG. 40. Mitsubishi pyrolysis heater.

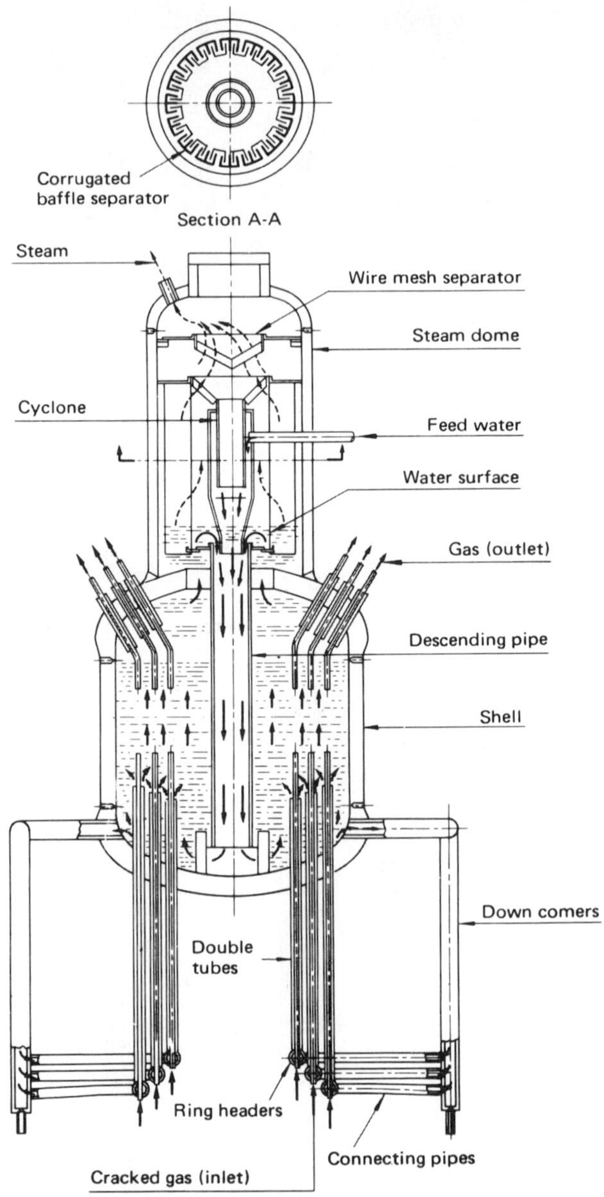

FIG. 41. Mitsubishi transferline exchanger.

is fired through terraced burners arranged at the top and midsection of the radiant box while feedstock enters through the convection section at the heater base.

The reactor coils are suspended in the usual manner from the top and consist of up and down passes so that partly concurrent- and partly counterflow conditions exist between hydrocarbon and flue gas flows. Burners are of the combination type that can fire fuel oil and gas simultaneously. The reactor coils consist of oval-shaped tubes.

For the same mass flow, elliptical tubes of equal perimeter as circular tubes will result in somewhat lower residence times (16% less for an axis ratio of 2:1) and a somewhat greater ratio of their perimeter will be exposed to reradiation from the incandescent walls of the combustion chamber. However, the tubes are more difficult to manufacture and might tend toward a peripherally more disuniform coke laydown. Comparisons of the performance of oval-shaped vs circular tubes for otherwise identical coil configurations have not been published.

In the Mitsubishi heater, two coils are connected to a transferline exchanger of rather unique design. Residence times from 0.3 to 0.6 s and capacities from 50,000 t/yr for full range naphthas and 40,000 t/yr for heavy gas oil have been quoted. It is also claimed that the M-TCF shows improved thermal efficiencies and simplifies the decoking procedure while the heater is in service [173, 174].

8.6.3. Transferline Exchangers

Transferline exchangers, also referred to in the literature as "quench coolers," have the purpose of rapidly cooling the pyrolysis coil effluents in a time interval measured in milliseconds and thereby "freezing" the reactor outlet composition at its momentary value to prevent degradation of the olefin yield through continuing secondary reactions. The cooling effect is most intense at the transferline exchanger inlet where the temperature difference between gas and coolant is largest. Yet cooling must stop at a sufficiently high temperature to minimize the condensation of the heaviest fuel oil fractions in the coil effluent. Under these circumstances the most efficient cooling effect is obtained by the vaporization of highly demineralized water into steam at pressures of ~12.5 MPa. Such steam may subsequently be superheated in the convection section of the pyrolysis heater before it is used in compressor drivers. The heater steam system thus consists of transferline exchangers, a steam drum, and the necessary piping and instrumentation on the gas and steam side. Gas flow inside the tubes as well as steam side. Gas flow inside the tubes as well as steam-feedwater flow outside is in an upward direction. The steam–water mixture issuing from the exchanger is separated in an elevated drum, the water returning to the inlet side of the exchanger by the thermosyphon effect at a rate of about 10 kg of water per kg of steam produced.

There are a number of difficult problems attending the design of these exchangers. Usually two reactor coils through a crossover fitting are connected to one exchanger. The crossover fitting in the form of a Y-shaped piece is installed outside of the radiant box and its volume is accounted for as part of the coil volume when calculating residence time. The inlet cone (of minimum volume) to the exchanger, directly flanged to the crossover fitting where it penetrates the roof of the radiant box, distributes the gas flow evenly over the exchanger inlet tube sheet. Measurements have shown that flow to the tubes varies by not over ±5.0% from the average value. Greater variations would cause an unbalance in flow and heat flux rates that would impair performance.

It is the designer's aim to produce an exchanger whose run length is at least equal and preferably longer than that of the heater coils. Toward this end the exchangers must be dimensioned to cool the gases to 400 to 500°C in naphtha service and 500 to over 600°C in gas oil service, the lower values corresponding to the exchanger when in clean condition.

Gradual impairment of the effectiveness of the heat transfer surface, accompanied by a rise in outlet temperature, is nonetheless unavoidable and requires periodic cleaning.

Heat transfer in the main body of the exchanger involves nucleate and film boiling. Heat flux rates across and in the immediate vicinity of the inlet tube sheet are in the order of 400 kW/m^2 in two of the most widely used exchanger types but taper off logarithmically along the tube to average in the order of 60 kW/m^2 for an element of 4.0 to 5.0 m in length between inner faces of tube sheets. These average flux rates are well below values at which film boiling becomes the dominant regime. Inlet tube sheets are rather thin (due to a differential expansion problem) and need staying which is done differently in different designs. Tube diameters range from 25 to 50 mm. Uniform inlet distribution is more difficult to achieve in large bundles. For prefabricated, commercially available models, material specifications are usually the following: Tube sheets, tubes, and cones of carbon steel (1.25% Cr, 0.5% Mo); shell, steam drum, and piping to accord with codes recognized in the country of erection. Inlet and outlet cones are generally internally form-clad with a hard surfaced asbestos-cement for thermal insulation.

For continuous, maintenance-free operation of steam producers in such critical very high temperature service, the quality of the feedwater is of paramount importance. Malperformance of the steam system has, in fact, in most instances been traced to substandard quality feedwater.

The feedwater has to be fully dimineralized by one of the proven methods, should have a pH value of 9.0 to 9.5, and must not contain any particulate matter, although means for venting and blowing down while in service are provided in some of the exchanger designs.

At highly elevated temperatures, water reacts to a measurable extent with iron to the magnetite form, $Fe_3O_4(Fe_2O_3 \cdot FeO)$ of iron oxide. A magnetite film which forms on the tube walls inhibits further oxidation. The film may be disturbed by excessive stresses induced by differences in thermal expansion and by other factors, if not considered in the design. Loss of metal through stress corrosion and lack of film protection mean a greatly shortened service life of the exchanger [175].

Notable for their widespread use in commercial installation are the following types of exchangers.

Borsig, Berlin, Germany, produces exchangers to the clients' specifications. The Borsig design (Fig. 42) is distinguished by an equidistant square tube layout, the tubes being fully immersed in boiling water. Feedwater is directed through a circumferential shroud against the inlet tube sheet, which is stayed by internal crossbars. Tubes are welded in place. The inlet cone may contain concentric, conically shaped directional baffles to help achieve uniform gas distribution over the inlet tube sheet.

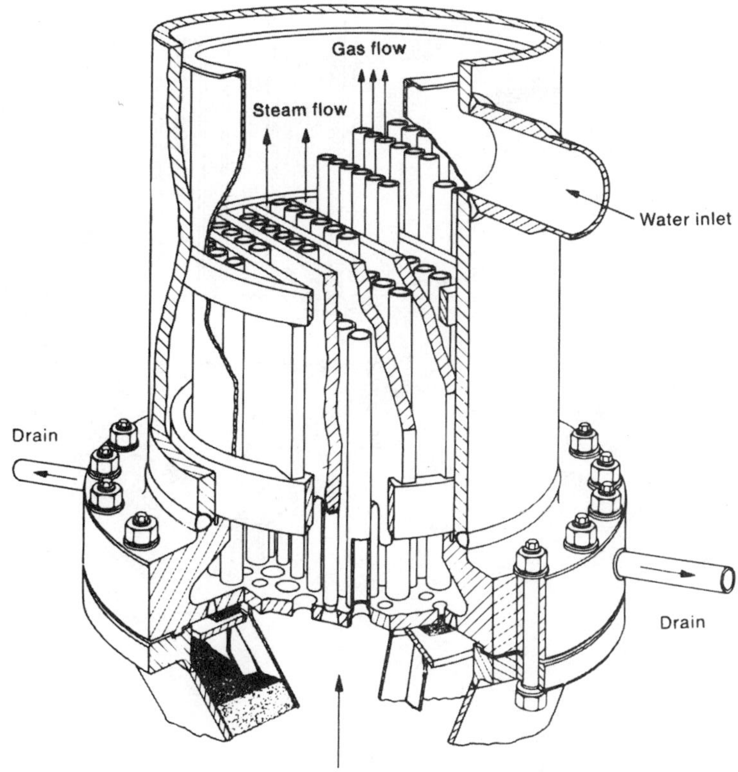

FIG. 42. Borsig transferline exchanger.

Schmidtsche Heissdampf G.m.b.H., Kassel, Germany, offers a design also featuring an equidistant square tube layout. However, each tube is surrounded by a concentric outer tube which carries the water–steam mixture (Fig. 43). The annuli between inner and outer tubes are supplied with feedwater over horizontal, oval-shaped headers that are in a rather unique way welded together to form the tube sheets. Thus the tubes themselves act as stays for the tube sheets. Sufficient hydraulic head is relied upon to assist in obtaining equal distribution of the feedwater flow to the annullar risers over the oval-shaped lateral headers.

Mitsubishi Petrochemical Co. and Mitsubishi Heavy Industries, both of Tokyo, Japan, offer a transferline exchanger design with integral steam drum and cyclones for the steam–water separation. The individual pyrolysis gas tubes are located in a pool of boiler water in the form of a half helix to allow for thermal expansion during the combustion of deposits. To avoid thermal stresses in the shell of the steam drum, a short double tube arrangement is used for installing the transfer surface through the shell. Boiler feedwater is supplied to the annuli through thermosyphon action. This design largely eliminates the need for mechanical cleaning and, it is claimed, accomplishes the regenera-

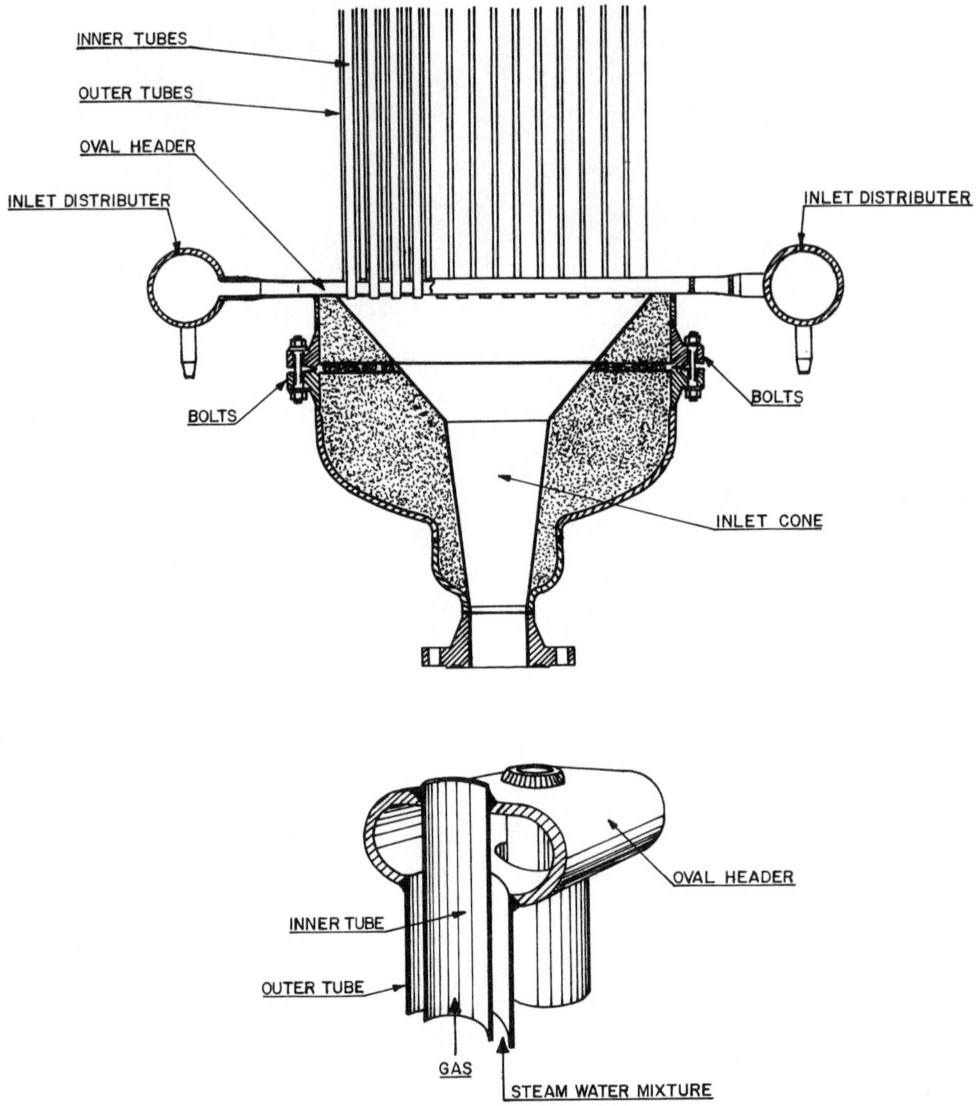

FIG. 43. SHG transferline exchanger.

tion of both pyrolysis furnaces and associated transferline exchangers in one operation [173, 174].

Mitsui Petrochemical Industries, Tokyo, Japan, has developed an "innovative quench cooler" that uses three concentric tubes as the tube element (Fig. 44). The innermost tube in downflow provides feedwater from an elevated reservoir to the inner annulus wherein vaporization takes place during upflow. The heater effluent in upflow passes through the second, outer annulus. This design has the obvious advantage that it is free from constraints due to differences in thermal expansion of the tubes. This design requires and incorporates a procedure for removal of deposits by combustion. Mechanical cleaning of the outer annulus would appear almost impossible. For further details of this design the reader is referred to Ref. 77.

Other types of quench coolers are in operation, among them proprietary models developed by Stone & Webster, that have been briefly referred to in connection with the special pyrolysis heaters developed by these firms. In the Stone & Webster arrangement, two U-tube coils are reportedly connected to

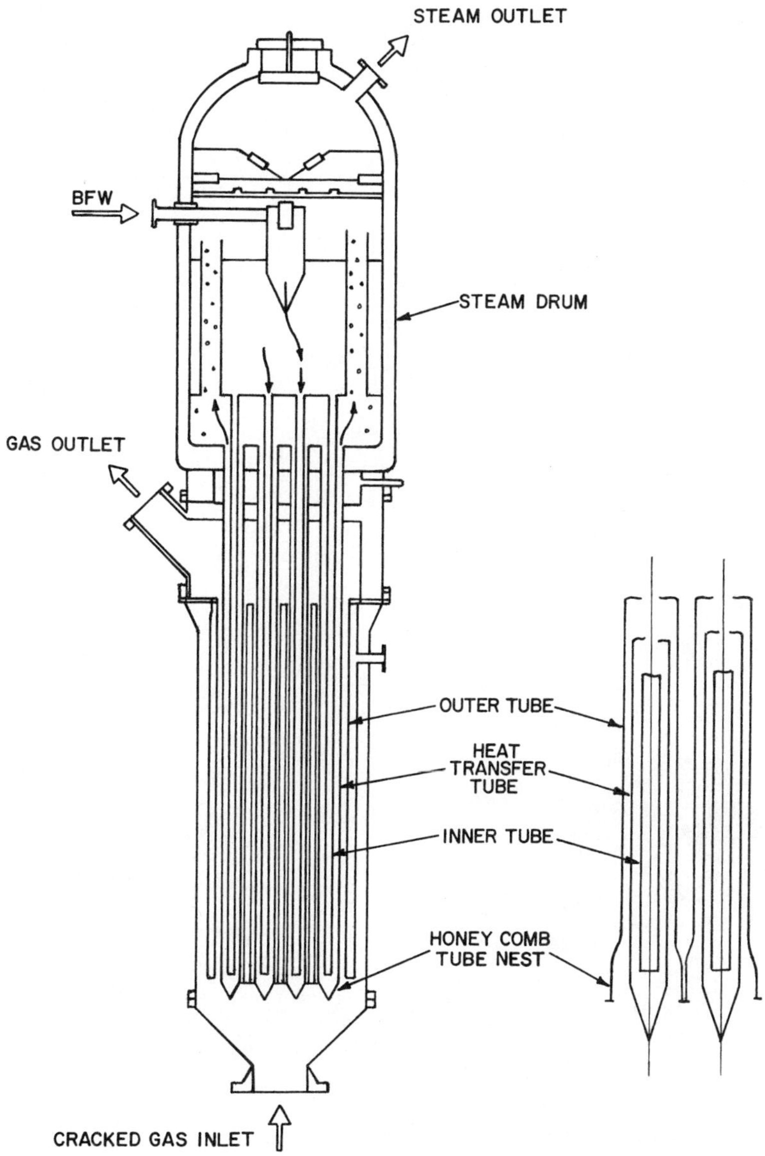

FIG. 44. Mitsui "quencher cooler."

one vertical transferline exchanger. A recent Japanese proposition by Babcock-Hitachi K.K. of two series connected quench coolers, of which specimens have been installed in a Japanese olefins producer and of which a fragmentary outline is given in Ref. 176, also deserves mention. In this design, which employs small diameter tubes for the reactor coils, each coil outlet is connected to a single tube exchanger dimensioned to cool the effluent to 650°C. At this point the flows from these exchangers are combined by manifolding to pass through a single element for further cooling to 400°C and additional steam generation. Both sets of exchangers receive boiler feedwater from the same steam drum. The proposed system of quenching shows considerable similarity with the one usually installed by Stone & Webster.

9. Ethylene Plants and Petroleum Refineries

An ethylene plant with a capacity of 500,000 t/yr designed to process petroleum fractions requires in the range of 1.5 to 2.0 million t/yr of feedstock which represents a substantial portion of the total production of even a large petroleum refinery. While it is possible to secure such large quantities of feedstock in the open market, more often than not large ethylene plants are built in close proximity to refineries with considerable synergism between both facilities. The dependence on a secure feedstock source and the considerable production of fuel-type by-products of ethylene plants processing heavy feedstocks would seem to promote a close relationship between ethylene plants and oil refineries. In the United States more than 8% of the 5.7 million tons of annual ethylene production capacity scheduled to be added between 1977 and 1980 will be owned by oil companies or their petrochemical subsidiaries [18]. Some capacity will be added by joint ventures of petroleum and chemical companies, a concept which has proven successful in Europe and Japan.

For the refiner an ethylene plant will be another conversion unit such as fluid catalytic crackers, hydrocrackers, or catalytic reformers. The conversion of various petroleum fractions into high-value petrochemicals is economically attractive to the refiner, not only because the demand growth for petrochemicals is expected to exceed the demand growth for the traditional fuel products of refineries, but also because basic petrochemicals are less likely to be subjected to governmental controls.

Carrying the conversion concept further, Gambro et al. [177] introduced the petrochemical refinery concept for the maximum conversion of crude oil to petrochemicals. Stock et al. [178] studied configurations of petrochemical refineries with increasing ratios of petrochemical to fuel production and shows that up to 26% ethylene yield and close to 60% total petrochemical yield from crude oil can be achieved with proven refining and ethylene technology. The yields of ethylene, propylene, butadiene, and aromatics make up the potential petrochemicals yield. Such high yields of petrochemicals require severe hydrotreating of the heavier petroleum fractions.

Sinkar [179] reported on a petrochemical refinery concept incorporating a fluid catalytic cracker for the atmospheric residuum (HOC Process) with high yields of total petrochemicals of which, however, the aromatics represent a larger portion than in the case of extensive hydrotreating.

While the substantial conversion of crude oil to petrochemicals, and to ethylene specifically, is technically feasible, the economically most attractive petrochemical refinery configuration provides for the production of both petrochemicals and traditional refinery products. Ethylene plants associated with refineries, if designed for feedstock flexibility, can be supplied with petroleum fractions for which a variable seasonal demand exists in the fuels market. Ethylene demand exhibits little or no seasonal fluctuation.

Traditionally, ethylene plants have preferred hydrocarbon feedstocks for which the lowest competing demand existed in the fuels market and which were thus available at relatively lower cost. In Europe, for example, naphtha was valued for many years as approximately equal to residual fuel oil and much lower than middle distillates. Since the early to mid-1970s, much of this has changed. Middle distillates are widely used as ethylene plant feedstocks, and there is considerable interest in vacuum gas oils as raw materials for ethylene production. Virgin vacuum gas oils from most crudes produce relatively low yields of olefins. In the pyrolysis of vacuum gas oils, the production of unstable pyrolysis fuel oil may exceed the ethylene production by more than 50%, and the generally high sulfur content of gas oils results in high sulfur fuel products and causes operational problems in ethylene plants. All of this is a consequence of the high content of polycyclic naphthenes and aromatics in gas oils, exceeding 50 mass% in vacuum gas oils. These components have little value as precursors of light olefins because of their already low hydrogen to carbon ratio. Sulfur is bound in these materials in a variety of structures.

To overcome these adversities in the processing of gas oils in ethylene plants, some form of feed pretreatment is generally required. As a minimum, gas oil feedstocks are usually hydrodesulfurized to a maximum sulfur content of 0.2 mass%. Hydrodesulfurization, a well-established refining process, not only converts sulfur to hydrogen sulfide but also effects the substitution of hydrogen for sulfur in the molecular structure of the feedstock components with an attendant decrease in density and increase in the hydrogen to carbon ratio. As a result, the ethylene yield potential of the hydrotreated material is enhanced over that of the virgin material.

For several years development work has been in progress on methods of more severe hydrotreatment to convert the polycyclic compounds and, in particular, the condensed aromatics of gas oils into compounds more suitable as pyrolysis feedstocks. Hydrocracking to a naphtha-type material, widely practiced in refineries, does not represent an attractive approach because naphtha-type hydrocrackates are highly isoparaffinic and yield but little ethylene through pyrolysis. However, nearly complete saturation of the aromatic compounds with minimum cracking to lighter materials achieves a very significant enhancement of the pyrolysis yields from gas oils. Sze et al. [180] and Davis et al. [181] reported pyrolysis yields of 26 to 29 mass% ethylene after hydrotreating gas oils in the presence of nickel–tungsten or cobalt–molybdenum catalysts at temperatures of approximately 350°C and under pressures of 7.0 to 17.0 MPa. The reported pyrolysis yield spectra are comparable to yields of typical full-range virgin naphthas. Thus this hydrotreating technology, once brought to industrial maturity, may represent a viable route to petrochemicals from vacuum gas oils for both new production facilities and existing naphtha-based ethylene plants. For the refining industry, which is accustomed to fluid catalytic cracking and hydrocracking, this technology represents another option for the conversion of vacuum gas oils to more valuable products.

Aromatics can also be removed from gas oils by solvent extraction, but this process yields an aromatics-rich stream of little value. Zdonik et al. [182] have investigated various gas oil pretreatment options including hydrodesulfurization, standard hydrocracking, and solvent extraction of aromatics. All of these result in a certain improvement of the petrochemicals yield potential of the pyrolysis feedstock. The yield data for the hydrocracking option show successively lower ethylene yields for the lighter hydrocracked product, indicating that the formation of lighter materials should be minimized to obtain maximum feedstock improvement.

Pyrolysis feedstock pretreatment represents one important link between petroleum refineries and ethylene plants also because the latter is a source of hydrogen. A further degree of integration of both facilities is economically advantageous if production of benzene and alkylbenzenes is desired in quantities requiring both catalytic reforming and aromatics recovery from pyrogasoline.

10. The Economics of Ethylene Manufacture

The raw materials for the production of ethylene are either natural gas components or petroleum fractions. Olefins are the desired products. Pyrolysis gasoline and fuel oil are sold in the fuels market. The economics of ethylene manufacture thus is strongly influenced by forces affecting the markets of natural gas and petroleum products.

Figure 45 shows the prices of ethylene, propylene, and butadiene in the United States over the last 20 years. Until the early 1970s the ethylene price declined as a result of cost reductions achieved with steadily more efficient plants of increasingly larger capacity. The price history of aromatics and fuel oil shown in Fig. 46 is quite similar to that of ethylene. The steep rise of all prices beginning in 1973 was initiated by the action of the OPEC countries to raise crude oil prices. The effect of these developments on the prices of primary petroleum products in Northern Europe is shown in Fig. 47.

The years following 1973 were also marked by a severe escalation of costs in the capital goods market that notably increased the investment cost of new ethylene production facilities. Both of these factors combined to raise the price of ethylene to new levels. In 1976, as feedstock costs stabilized, the price of ethylene began to level off, albeit at a record high level.

With continuing inflation, a new production facility, consuming less raw materials and energy per unit of product, may not compete with or cause the retirement of less efficient but fully depreciated older facilities.

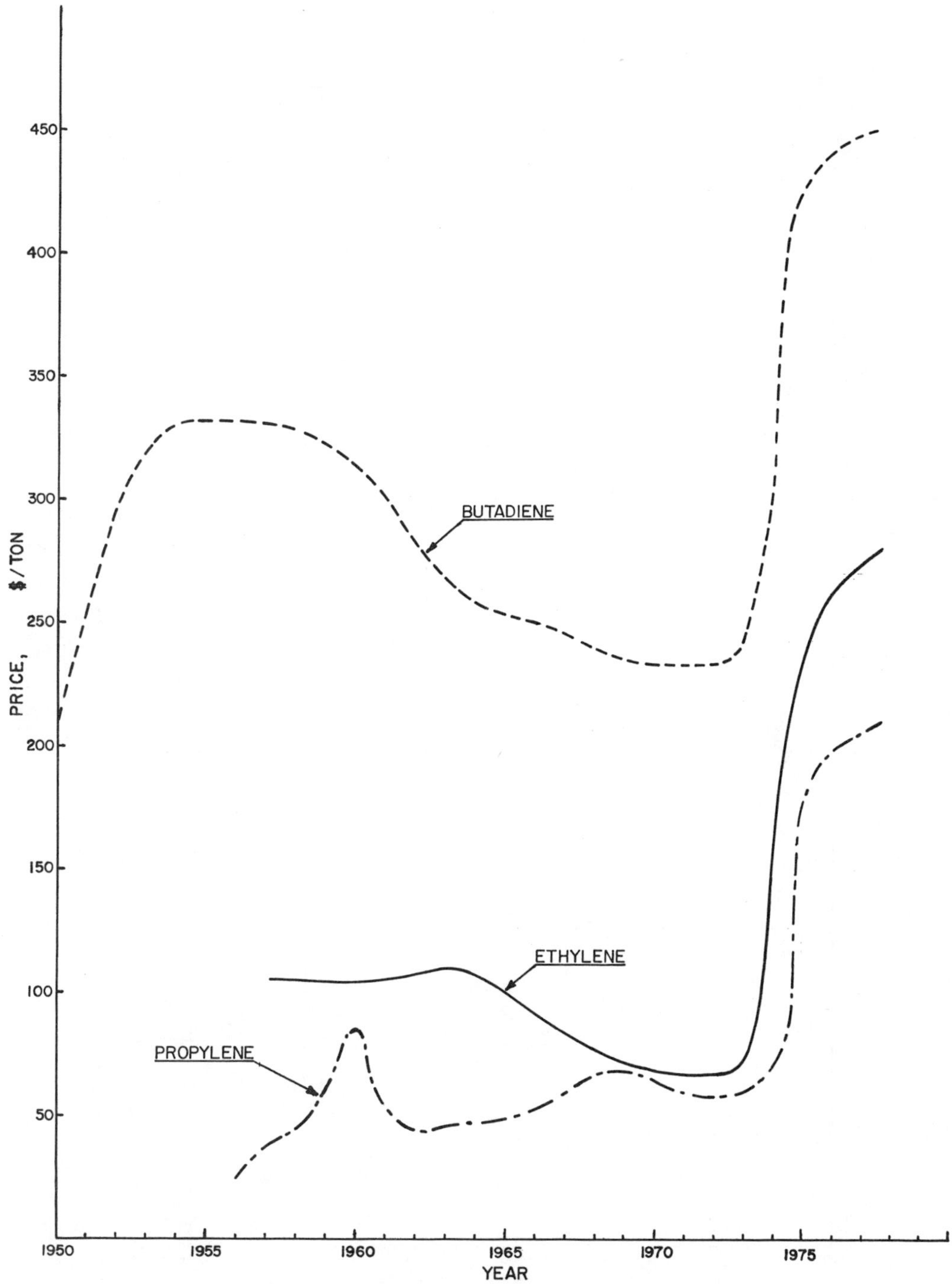

FIG. 45. Historic price trend of butadiene, ethylene, and propylene in the United States.

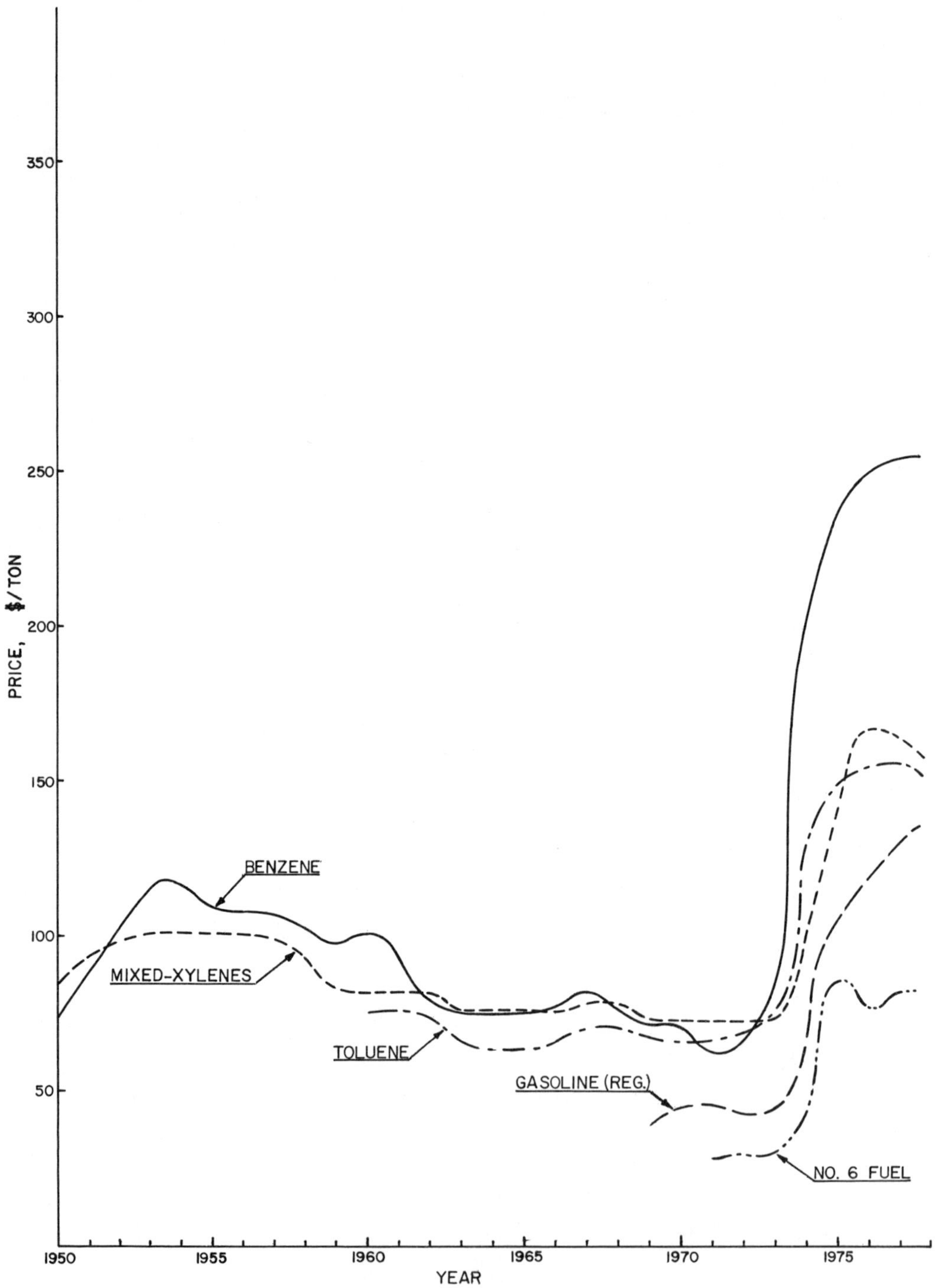

FIG. 46. Historic price trend of benzene, toluene, mixed xylene, gasoline (regular), and No. 6 fuel oil in the United States.

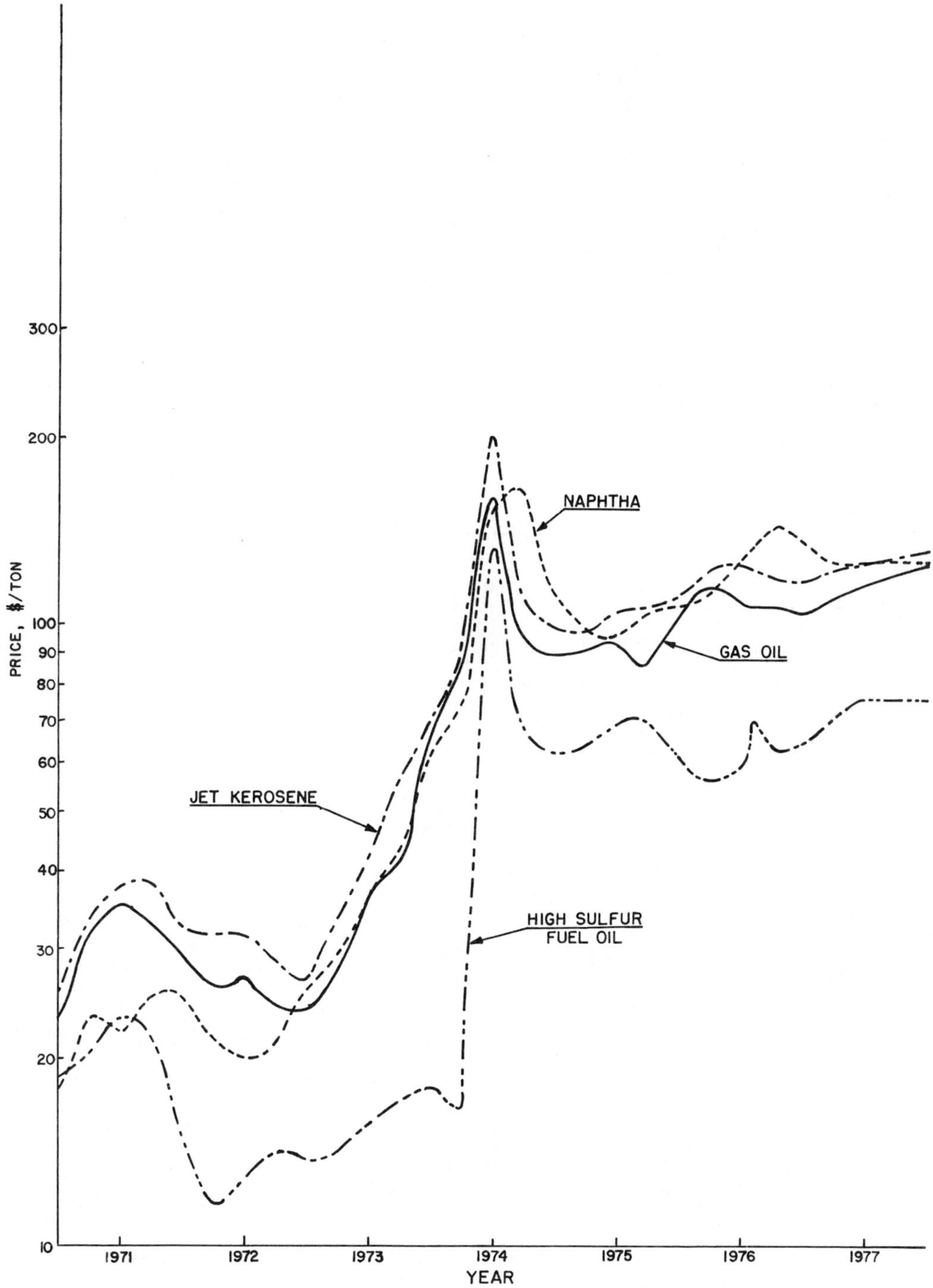

FIG. 47. Historic price trend of naphtha, jet kerosene, gas oil, and high sulfur fuel oil in Northern Europe.

An economic evaluation of an ethylene manufacturing facility has to take cognizance of the fact that the monetary values of feedstocks and products are, to a considerable degree, interrelated with the values of hydrocarbon materials in the fuel market. The typical ethylene plant, unless solely fed with ethane, produces by-products in significant volumes. The combined by-products volume of plants processing naphtha and heavier feedstocks, for example, exceeds the production rate of ethylene by more than a factor of 2. A meaningful economic analysis will have to take into account the interrelationship of hydrocarbon values. Sensitivity analyses of ethylene manufacturing costs for large variations in feedstock costs based on the assumption of constant values for the by-products, while at times reported in the literature, are of limited utility for the assessment of the economic feasibility of an ethylene manufacturing facility.

10.1. Economic Analysis of Ethylene Manufacturing Facilities

Given the present uncertainties surrounding the markets for hydrocarbons and petrochemicals, an economic analysis is best based on a segregation of the costs and revenues of ethylene manufacture. By the simplest definition, economic feasibility requires that the sum of total revenues from all products of an ethylene production facility less the cost of raw materials is at least equal to or exceeds the cost of processing the raw materials in the facility. By definition, the cost of processing includes all operational expenses as well as capital-related charges.

In mathematical terms:

$$\sum_{1}^{i} Y_i V_i - V_f \geq C_p$$

where Y_i = net yield of the ith product leaving the plant as mass-fraction per unit feedstock
V_i = value of the ith product (\$/t)
V_f = value of the feedstock (\$/t)
C_p = cost of processing per unit mass of feedstock (\$/t)

The left-hand side of this inequality, often referred to as the "gross margin," contains the major cost elements that are market dependent.

The cost of raw materials and the market value of products at any one time are different at different locations over the globe; in Japan, for instance, the cost of naphtha is government controlled. Over the life of a plant, these costs may change considerably. Product yields will change relatively more slowly as technological advances are being made.

In accounting for total revenues, a deduction must be made for the fuel consumed by the plant. When the feedstock is ethane, the plant is not self-sufficient in fuel; additional fuel will usually have to be imported, the cost of which enters as a negative revenue.

The cost of processing, which does not include the cost of feedstock and fuel, is foremost a function of the charges that derive from the total capital cost of a facility. In turn, these charges depend on two parameters: the ethylene-producing capacity of the facility and the type of feedstock that it is designed to use. The cost of processing (C_p) is shown in Fig. 48 for a new ethylene facility on the U.S. Gulf Coast as a function of ethylene capacity and feedstock type. It clearly shows the cost advantage that larger facilities have due to the economy of scale. It also shows that processing costs referred to 1 t of feedstock drop rapidly the heavier the feedstock. It is to be emphasized that Fig. 48 represents costs referred to 1 t of feedstock. In a comparative case of two plants of the same ethylene capacity, one for a light and the second for a heavy feedstock, the total processing cost calculated on the basis of yields for the different volumes of chargestocks required will in most instances be greater for the heavier feedstock. However, the gross margin will also increase due to the extra revenue from the greater volume of by-products produced. The latter may more than compensate for the higher processing cost or it may fall short of doing so, depending on the difference in yield patterns as further discussed below.

10.2. Elements of the Cost of Processing

The cost of processing is made up of three different and distinct elements:

Charges related to the total capital investment of the plant
Operating costs, exclusive of fuel, that vary with the production rate
The cost of operating labor and supervision plus charges normally classified as overhead

Average industry values of these charges, further discussed in the following paragraphs, have been used in the development of the "cost of processing" data represented in Fig. 48.

10.2.1. Capital-Related Charges

The total capital cost of ethylene plants per annual ton of ethylene produced prevailing in late 1977 on the U.S. Gulf Coast, which is the cost reference location of the petrochemical industry most everywhere, is shown in Fig. 49 as a function of plant capacity for a variety of feedstocks ranging from ethane to vacuum gas oil.

The investment that is represented accounts for the cost of the Inside Battery Limits (ISBL) process facility as well as the cost of the Offsite Battery Limits (OSBL) support facilities. In Fig. 49 the cost of the OSBL facilities was taken as 50% of the cost of the ISBL plant. The figure may vary considerably due to different local conditions; the value of 50% represents an average for the industry and will typically include the cost of site preparation, utility systems,

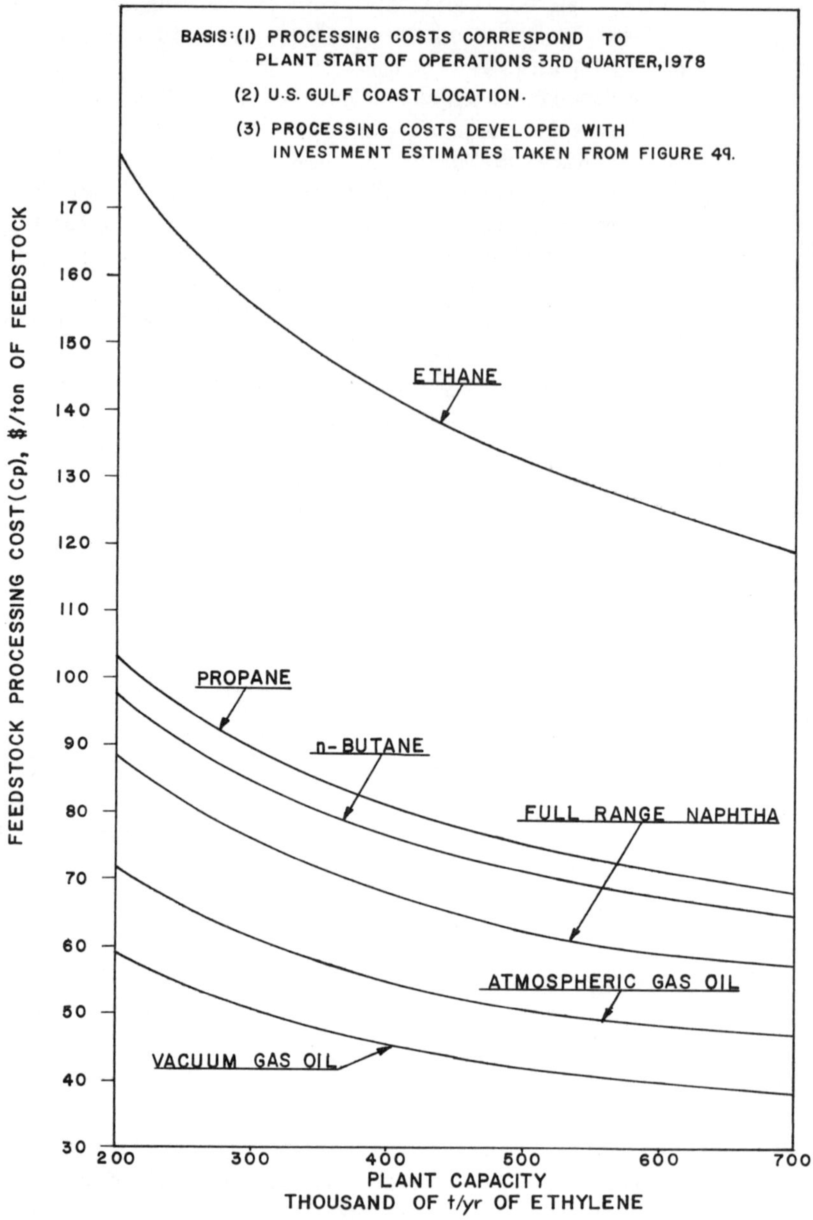

FIG. 48. Feedstock processing cost estimate as a function of feedstock and capacity.

storage facilities, warehouse, effluent treating and pollution abatement systems, loading and unloading facilities, roadways, a flare stack, and yard piping.

The ISBL portion of the total facility refers to single train process units, i.e., a plant with a single line of compressors and fractionators.

Figure 49 indicates that for a given ethylene capacity, the heavier the feedstock, the higher is the plant cost. It also illustrates the effect of scale on unit costs. By way of example, the total investment per ton of ethylene in a 200,000-t/yr naphtha based facility per Fig. 48 will be \$670 while for the 500,000-t/yr facility it will be only \$480.

The investment costs shown in Fig. 49 contain the fixed capital expenses required to establish an operating production facility. Included are contractor service and technology license fees. Excluded are land costs, start-up expenses, and contingencies. Working capital is not considered a component of fixed capital costs.

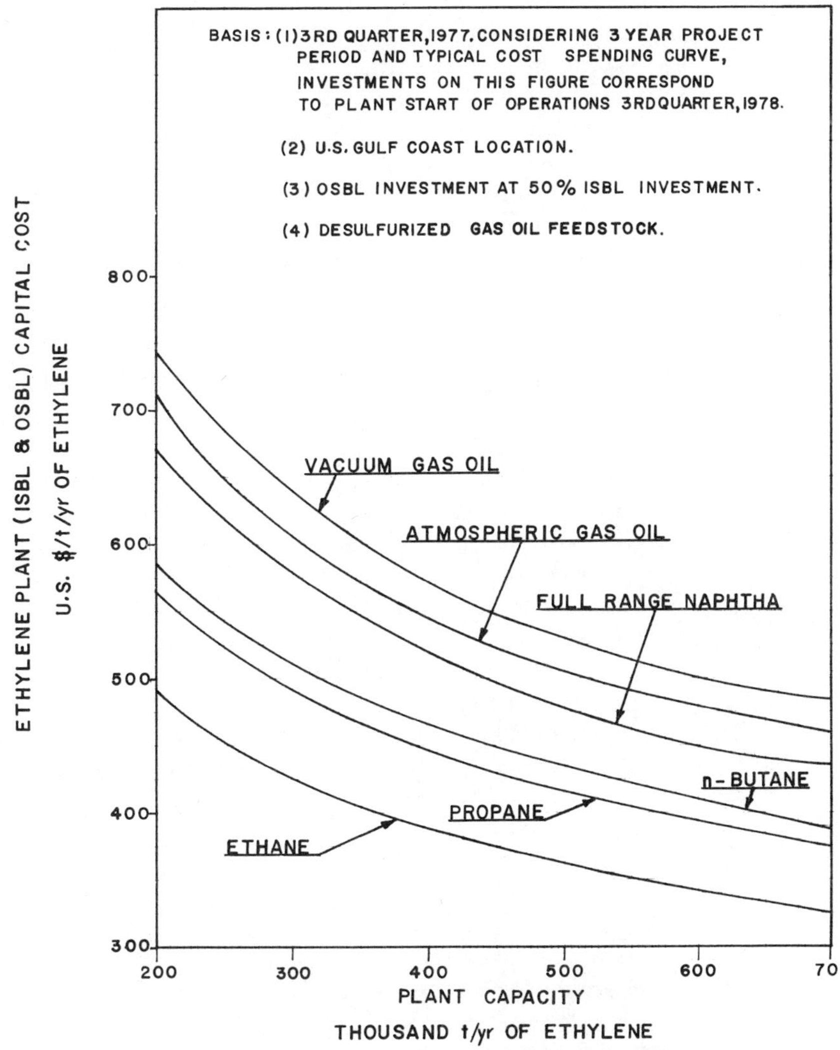

FIG. 49. Plant investment estimate as a function of feedstock and capacity.

The total plant investment gives rise to the following investment- or capital-related charges.

Depreciation represents the annual rate at which the plant investment is to be written off or retired as a result of wear, age, and obsolescence. For petrochemical facilities the allowable rate typically ranges from 7 to 12% for ISBL and from 4 to 7% for OSBL facilities.

Maintenance is another capital-related charge. Obviously, the plant must be maintained in operating order, whether or not it is operated. Permanently engaged maintenance crews must be on hand to service all parts of the plant. These charges nominally range from about 3 to 5% of the total investment annually, about half of it being for material replacements and half for maintenance labor.

Still other capital-related charges are insurance and property taxes. Insurance costs vary somewhat over the life of the plant and depend on whether the plant owner-operator is self-insured, i.e., he assumes the risks of losses due to accidents, or he contracts with insurance companies. Local property tax rate varies with location and tends to increase with time.

The cost of insurance and of taxes may nevertheless from experience be charged at a constant rate over the life of the plant and may thus be frozen into the total capital-related costs.

Insurance costs range in the order of 0.5 to 2% annually and taxes in the order of 1 to 1.5% annually of total investment.

Finally, the so-called "return on investment" represents the pretax profit or financial reward to the owner of the production facility to compensate him for the risks that he is undertaking.

The actual return on investment over the life of the plant will vary with the gross margin realized and the operating rate. The cash flow associated with the operation of a plant may vary substantially with time as is often the case with very large plants which may reach full capacity only several years after start-up. In these cases the discounted cash flow method of economic analysis, which takes into account the time-dependent value of money, will yield a more realistic picture of the profitability of a venture than would the return on investment based on full capacity utilization.

The return on investment is used as the measure of profitability in this section. The desired pretax annual rate of return for a facility producing commodity petrochemicals is typically in the range of 20 to 30% of the investment. The total of the annual capital-related charges as described above will thus be in the range of 30 to 50% of the investment cost.

10.2.2. Variable Operating Costs

Variable operating costs are processing expenses that vary with the rate of production; they comprise the cost of electric power, chemicals, catalysts, desiccants, fresh water, and steam to the extent that the latter is imported rather than internally generated. Variable operating costs represent only a relatively small fraction of the total cost of processing. They are typically in the range of 10 to 15% of the cost of processing.

10.2.3. Labor, Supervision, and Overhead Costs

The operation of an olefins plant requires cadres of skilled plant operators and technical supervisors. A typical naphtha based 500,000 t/yr ethylene plant requires approximately 15 to 20 operators per shift for the operation of the process facilities and the support systems.

Supervision and overhead costs cover a variety of expenses including the cost of the plant's management (both technical and administrative), technical, and laboratory services as well as corporate charges.

The cost of labor, supervision, and overhead is independent of the rate of production but will increase over the life of the plant as a result of salary and wage inflation. Supervision and overhead costs vary from company to company and are treated as confidential information. To arrive at the results shown in Fig. 48, they were assumed to equal 200% per manhour of operating labor including benefits, a figure that has been inferred by industry as of the proper order of magnitude.

10.3. Application of the Cost of Processing Method

The data and methodology presented in the previous paragraphs are considered useful in the preliminary planning of an ethylene manufacturing facility. It is realistic to assume that the cost of processing will increase proportionally with inflation of the cost of capital goods. Given a product yield pattern and the projected prices of products and feedstocks, the gross margin can be calculated and compared with the cost of processing shown in Fig. 49, escalated as necessary, to estimate the economic feasibility of a planned ethylene manufacturing facility.

As an example, the projected feedstock costs presented in Table 18 are assumed to have prevailed on the U.S. Gulf Coast in late 1978.

The product yields are assumed to be those listed in Tables 10 and 11 of Chapter 6, and revenues are suitably modified to account for the fuel requirements of the plant. The product values assumed for this example (Table 19) are projections to late 1978 from Figs. 45 and 46. Byproduct values were calculated on the value of the by-product components. For the example, the

TABLE 18 Feedstock Costs (U.S. Gulf Coast, late 1978)

Feedstock	Value ($/t)
Ethane	126
Propane	129
n-Butane	129
Full-range naphtha	135
Atmospheric gas oil	120
Vacuum gas oil	108

TABLE 19 Product Values (projected for U.S. Gulf Coast area as of late 1978)

Product	Value ($/t)
Ethylene	292
Propylene	218
C_4-product	229–355
Pyrogasoline	141–173
Fuel gas	157
Fuel oil	90

value of the C_4-stream is based on the amount of butadiene that it contains, less extraction costs plus fuel value of C_4-components other than butadiene. The value of the gasoline fraction in reference to the recovery of aromatics is calculated in a similar manner. The value ranges, where shown, do reflect the different yields obtained from different quality feedstocks.

Based on the projected costs of feedstocks and product values, the gross margin may be compared with the cost of processing feedstocks in, for instance, a 500,000-t/yr ethylene plant. Table 20 shows the results.

On the basis of the assumed product values and yields, these results indicate that vacuum gas oil seems presently to be the most attractive feedstock. Revenues could shrink by the equivalent of 15 $/t of feedstock or conversely the price of feedstock could rise by 15 $/t without this change having an effect on the stipulated return on investment. Propane would appear to be the least desirable feedstock. All others seem grouped together within a range that borders the estimated accuracy of this analysis so that conclusions can only be drawn on a case to case basis. To some minor degree this also holds for vacuum gas oil as the technology of producing olefins from vacuum gas oil has not as yet fully matured.

Extraneous factors not here considered, such as long-term feedstock availability, stable consumer contracts, the energy situation in general, and the

TABLE 20 Economic Feasibility of Ethylene Manufacture from Different Feedstocks (all values in dollar per ton of feedstock; (capacity 500,000 t/yr of ethylene)

Feedstock	Gross Margin (GM)	Cost of Processing (C_p)	GM − C_p
Ethane	134	133	1
Propane	68	76	−8
n-Butane	68	72	−4
Full-range naphtha	60	63	−3
Atmospheric gas oil	56	51	5
Vacuum gas oil	57	42	15

prevailing economic climate, may well have a more decisive influence on the selection of a particular feedstock.

10.4. Outlook

As of early 1978, it was estimated that the worldwide demand for ethylene over the near term would grow by an annual 8% until the early 1980s and then diminish to about 6% by 1985 [18]. At that time United States production may reach in the order of 26 million t/yr.

In the United States as of 1975, about 75% of the ethylene produced still came from ethane and LPG, with the remainder from petroleum fractions. The steady decline of natural gas supplies has helped to establish the present trend toward the use of heavier feedstocks such as naphthas, gas oils, and, conceivably in the not too distant future, even to full crudes.

The European petrochemical industry grew up on inexpensive naphtha since the oil industry there was oriented not so much toward the production of gasoline (15+% of the crude oil barrel vs 45% in the United States) as toward the production of gas oils and fuel oil for industrial purposes. An increasing demand for gasoline and the expected large volumes of natural gas from the North Sea fields that will be landed in Britain and Continental Europe have already started to reverse this trend. The prospects of different feedstocks for the petrochemical industry in Europe have recently been discussed by Pownell [185].

In the Middle East, Algeria, Indonesia, and elsewhere, extensive conservation measures, now being implemented, will bring to market by the mid-1980s large quantities, perhaps up to 30 million t/yr of condensates and LPG, which could foster the establishment of a native or joint foreign–native petrochemical industry in those countries. It might also bring about another reorientation in the feedstock situation elsewhere.

References

1. S. Angus et al., *Ethylene, 1972 International Thermodynamic Tables of Fluid State*, International Union of Pure and Applied Chemistry, Butterworths, London, 1972.
2. K. E. Starling and Y. C. Kwok, *Hydrocarbon Process.*, *50*(10), 90 (1971).
3. K. E. Starling, *Fluid Thermodynamic Properties for Light Hydrocarbon Systems*, Gulf Publishing Co., Houston, 1973.
4. *Industrial Needs for Critically Evaluated PVT Data of Ethylene and Related Substances, Report of Panel Discussions, June 13–16, 1972, Airlie House, Warnenton, Virginia*, Numerical Data Advisory Board, National Research Council, and Office of Standard Reference Data, National Bureau of Standards.
5. D. R. Douslin and R. H. Harrison, *J. Chem. Thermodyn.*, *8*, 301 (1976).
6. R. H. Harrison and D. R. Douslin, *J. Chem. Eng. Data*, *22*(1), 24 (1977).
7. J. R. Hastings and J. M. H. Levelt Sengers, *Equation-of-State Measurements for Ethylene Including the Critical Region*, Presented at AIChE 81st National Meeting, Kansas City, Missouri, April 11–14, 1976.
8. J. R. Hastings and J. M. H. Levelt Sengers, "Vapor Pressure, Critical Pressure, and Critical Isochore of Ethylene," in *Proceedings, Seventh Symposium on Thermophysical Properties*, American Society of Mechanical Engineers, New York, 1977, p. 794.
9. M. Waxman and H. A. Davis, *State of the Art-Determination of the Second Virial Coefficient of Ethylene* (Advances of Chemistry Series, No. 182), American Chemical Society, Washington, D.C., In Press.
10. F. S. Bonscher, L. M. Shipman, and L. C. Yen, *Hydrocarbon Process.*, *53*(1), 145 (1974).
11. K. -H. Imhausen et al., *Hydrocarbon Process.*, *55*(11), 155 (1976).
12. Anonymous, *Chem. Eng. News*, *55*(47), 6 (1977).
13. A. Hahn, A. Chaptal, and J. Sialelli, *Hydrocarbon Process.*, *54*(2), 89 (1975).
14. S. A. Miller, *Ethylene and Its Industrial Derivatives*, Baum, London, 1969, pp. 335 and 437.
15. Ref. 14, p. 413.
16. H. Riegel, H. D. Schindler, and M. C. Sze, *Chlorinated Hydrocarbons Produced via Transcat*, Presented at AIChE 74th meeting, New Orleans, Louisiana, March 14, 1973.
17. A. S. Nislick, H. Riegel, and M. A. Abrahams, *The Economics of Producing Chlorinated Hydrocarbons by the C-E Lummus Transcat Process*, Presented at AIChE 79th meeting, Houston, Texas, March 16–20, 1975.
18. A. J. Cahill, *Chem. Eng. Prog.*, *73*(7), 26 (1977).

19. M. Klaus, *Chem. Econ. Eng. Rev.*, *9*(4), 7 (1977).
20. F. Hori, *Chem. Econ. Eng. Rev.*, *9*(4), 17 (1977).
21. K. S. Wardale and G. Todd, *Chem. Eng.* (*London*), *213*, 247 (1967).
22. H. C. Schutt, *Oil Gas J.*, *56*(31), 74 (1958).
23. A. T. Zeiner, *Eur. Chem. News*, *15*, Storage and Distribution Survey (1969).
24. W. B. Clapp and L. F. Litzinger, *Pipeline Gas J.*, *198*(7), 72 (1971).
25. Anonymous, *Pet. Times*, *79*(2015), 22 (1975).
26. E. R. Ellis, *Engineer*, p. 264 (February 8, 1963).
27. Chicago Bridge & Iron Co., Private communication to L. Kniel, April 1977.
28. U.S. Department of Commerce, *Federal Energy Administration Report PB 255 948*, July 1976.
29. H. R. Hardy, *Proc.*, *Am. Gas Assoc.*, *76-T-56*, T350 (1976).
30. W. H. Litchfield et al., *Chem. Eng. Prog.*, *55*(4), 68 (1959).
31. Anonymous, *Eur. Chem. News*, *16*, Storage and Distribution Survey (1970).
32. Anonymous, *Rev.*, *Assoc. Fr. Technicians Pet.*, *202*, 25 (1970).
33. R. R. Quirio and R. R. Hultin, *Ethylene Storage near Midland, Michigan*, Presented at Symposium on Salt, North Ohio Geological Society, Cleveland, Ohio, 1963.
34. D. F. McCarthy, *Pipeline Gas J.*, *199*(3), 52 (1972).
35. K. C. Campbell, Phillips Petroleum Co., Private communication to L. Kniel, April 1977.
36. United Nations Conference, *Ocean Transport of Ethylene*, September 1964; also R. Boudet and M. H. Gertz, *World Pet.*, *36*(10), 58 (1965).
37. J. L. Whyte, *Oil Gas J.*, *66*(34), 68 (1968).
38. Anonymous, *Eur. Chem. News*, *25*(641), 10 (1974).
39. Anonymous, *Eur. Chem. News*, *25*, Storage and Distribution Survey (1974).
40. L. Kniel and F. Fussman (C-E Lummus), U.S. Patent 3,877,240 (April 15, 1975).
41. A. L. Waddams and G. Cann, *Hydrocarbon Process.*, *53*(11), 135 (1974).
42. K. Ito, *Chem. Econ. Eng. Rev.*, *6*(2), 19 (1974).
43. R. J. Lackey, "The IMCO Code for Gas Tankers: A Review of the Finalized Code," in *Proceedings*, *Gastech 75 LNG & LPG Technology Congress*, *Paris*, *France*, *September 30–October 3*, *1975*, p. 65.
44. Anonymous, *Eur. Chem. News*, *18*, Storage and Distribution Survey (1972).
45. C-E Lummus, Unpublished Data, 1973.
46. A. Sike, *Maritime Gas Transport in the Development of the World Chemical Industry*, Presented at Congres Franco-Americaine de Chimie Industrielle, Valley Forge, Pennsylvania, October 18–21, 1976.
47. Anonymous, "The Esso Ethylene Line," *Fluid Handl.*, (January 1962).
48. W. Von der Linde, *Oel Z.*, p. 32 (February 1975).
49. H. W. von Blumencron and H. Steinrotter, *Chem.-Ing.-Tech.*, *41*(13), 750 (1969).
50. R. Jozon, *Oil Gas Int.*, *10*(6), 109 (1970).
51. H. G. B. Nolan, *Oil Gas Int.*, *9*(6), 82 (1969).
52. H. G. B. Nolan, *Oil Gas Int.*, *9*(7), 38 (1969).
53. Anonymous, *Eur. Chem. News*, *27*(688), 6 (1975).
54. J. P. O'Donnell, *Oil Gas J.*, *69*(6), 54 (1971).
55. *Properties and Essential Information for Safe Handling and Use of Ethylene* (Chemical Safety Data Sheet SD-100), Manufacturing Chemists Association, Washington, D.C., 1973.
56. H. Hoefermann et al., *Chem. Ind.*, *21*, 860 (1969).
57. E. Mosberger, *Chem.-Ing.-Tech.*, *43*(3), 131 (1971).
58. H. G. Davis and R. G. Keister, *The Advanced Cracking Reactor* (*ACR*): *A Process for Cracking Hydrocarbon Liquids at Short Residence Times*, *High Temperatures*,

and Low Partial Pressure, Presented at ACS Meeting, Philadelphia, Pennsylvania, April 6–11, 1975.
59. *Specifications and Test Methods for Liquid Petroleum Gas*, Natural Gas Processors Association Publication, 2140–62.
60. O. Winter and M. T. Eng, *Hydrocarbon Process.*, *55*(11), 125 (1976).
61. K. Stock, J. Hanisian, and J. Bac, *Hydrocarbon Process.*, *55*(11), 151 (1976).
62. K. Stork, C. H. Viens, and P. C. Yeung, *Oil Gas J.*, *75*(35), 442 (1977).
63. A. M. Brownstein, *Oil Gas J.*, *75*(35), 431 (1977).
64. K. Hayashi, M. Nakaniwa, M. C. Sze, and A. A. Simone, *Erdoel Kohle*, *30*(2), 65 (1977).
65. A. W. Pratt and N. L. Foskett, *AIChE Trans.*, *42*, 149 (1946).
66. W. Wolf, *Chem. Eng. Prog.*, *72*(7), 53 (1976).
67. C. W. Wolf, D. W. Weiler, and E. G. Ragi, *Oil Gas J.*, *73*(35), 85 (1975).
68. R. R. Hudgins, "Catalytic Dehydration of Ethanol Using a Frequency Response Technique," Dissertation, Princeton University, June 1964.
69. M. E. Winfield, *Catalysts*, Vol. 7, Reinhold, New York, 1960, pp. 93–183.
70. L. Kniel, *Utilization of the LNG Cold Potential with Particular Reference to the Manufacture of Ethylene*, Presented at International Conference on Cryogenics in Fuel and Power Technology, London, England, March 28, 1969.
71. A. L. M. Van Eijnatten, *Chem. Eng. Prog.*, *73*(9), 69 (1977).
72. Idemitsu Petrochemical (K.K.) Tokuyama Plant, *Report of Accident Investigation Committee, July 28, 1973.*
73. G. F. Fryling, *Combustion Engineering*, Combustion Engineering, Inc., New York, 1966, pp. 8–17.
74. L. Verde et al., *Hydrocarbon Process.*, *56*(7), 102 (1977).
75. D. Sohns and C. Fuge, *Linde Rep. Sci. Technol.*, *23*, 11 (1976).
76. T. Sato and Y. Ohnishi, *Bull. Jpn. Pet. Inst.*, *13*(2), 279 (1971).
77. K. Sato, *Chem. Econ. Eng. Rev.*, *9*(6), 16 (1977).
78. J. Chen and M. J. Maddock, *Hydrocarbon Process.*, *52*(5), 147 (1973).
79. M. Picciotti, *Hydrocarbon Process.*, *56*(5), 183 (1977).
80. M. Picciotti, *Hydrocarbon Process.*, *56*(6), 163 (1977); *56*(9), 179 (1977).
81. M. Raab, *Linde Rep. Sci. Technol.*, *23*, 23 (1976).
82. K. Stork (C-E Lummus), U.S. Patent 3,536,610 (October 27, 1970).
83. K. Stork and K. Wagner, *Oil Gas J.*, *69*(10), 60 (1971).
84. M. Ruhemann, *The Separation of Gases*, 2nd ed., Clarendon, Oxford, England, 1949, Chap. 12.
85. W. K. Lam, G. I. Fischer, and G. M. Clancy, *Oil Gas J.*, *68*(20), 111 (1970).
86. K. Stork, L. Kniel, and S. Markbreiter (C-E Lummus), U.S. Patent 3,816,976 (June 18, 1974).
87. U. Lorber, H. Reimann, and F. Rottmayr, *Chem. Eng.*, *78*(17), 83 (1971)
88. F. M. Kuchta, I. Spolan, and M. G. Zabetakis, *Chem. Eng. Data*, *9*, 467 (1964).
89. K. C. Chao and J. D. Seader, *AIChE J.*, *7*, 598 (1961).
90. M. Benedict, G. B. Webb, and J. Rubin, *Chem. Phys.*, *8*, 334 (1940).
91. O. Redlich and J. N. W. Kwong, *Chem. Rev.*, *44*, 233 (1949).
92. R. F. Curl, Jr., and K. S. Pitzer, *Ind. Eng. Chem.*, *50*, 265 (1958).
93. L. C. Yen and R. E. Alexander, *AIChE J.*, *11*(2), 334 (1965).
94. E. Mosberger, *Chem.-Ing.-Tech.*, *43*(3), 131 (1971).
95. F. O. Rice and K. F. Herzfeld, *J. Am. Chem. Soc.*, *56*, 284 (1934).
96. F. O. Rice and K. K. Rice, *The Aliphatic Free Radicals*, John Hopkins Press, Baltimore, Maryland, 1935.
97. S. B. Zdonik, E. J. Green and L. P. Hallee, "Manufacturing Ethylene" reprinted from the *Oil & Gas Journal* (1966–1970), Petroleum Publishing Co., Tulsa,

Oklahoma, 1970.
98. H. C. Schutt, *Chem. Eng. Prog.*, *55*(1), 68 (1959).
99. R. E. Kinney and D. J. Crowley, *Ind. Eng. Chem.*, *46*, 258 (1954).
100. A. G. Volkan and G. C. April, *Pyrolysis of Propane and Butane—A Review*, Presented at AIChE 83rd National Meeting, Houston, Texas, March 23, 1977; also *Ind. Eng. Chem., Process Des. Dev.*, *16*(4), 429 (1977).
101. D. Edelson and D. L. Allara, *AIChE J.*, *19*(3), 638 (1973).
102. V. Illes, *Acta Chim. Acad. Sci. Hung.*, *59*(1), 35 (1969).
103. V. Illes, *Acta Chim. Acad. Sci. Hung.*, *59*(2), 229 (1969).
104. J. H. Purnell and C. P. Quinn, *Proc. R. Soc. London*, *270*, 267 (1962).
105. J. E. Blakemore, J. R. Barker, and W. H. Corcoran, *Ind. Eng. Chem., Fundam.*, *12*(2), 147 (1973).
106. A. G. Buekens and G. F. Froment, *Ind. Eng. Chem., Process Des. Dev.*, *10*(3), 309 (1971).
107. D. Kunzru, Y. T. Shah, and E. B. Stuart, *Ind. Eng. Chem., Process Des. Dev.*, *12*(3), 339 (1973).
108. S. S. Levuch, A. Z. Savchenkov, S. S. Abadzhev, and V. U. Shevchuk, *Neftekhimiya*, *10*(5), 656 (1970).
109. P. A. Longwell and B. H. Sage, *J. Chem. Eng. Data*, *5*, 322 (1960).
110. V. V. Babash, T. N. Mukhina, and M. E. Aerov, *Int. Chem. Eng.*, *15*(2), 368 (1975).
111. V. Illes, I. Pleszkats, and L. Szepesy, *Erdoel Kohle, Erdgas, Petrochem.*, *22*(4), 201 (1969).
112. Y. T. Shah, E. B. Stuart, and D. Kunzru, *Ind. Eng. Chem., Process Des. Dev.*, *12*(3), 344 (1973).
113. D. Kunzru, Y. T. Shah, and E. B. Stuart, *Ind. Eng. Chem., Process Des. Dev.*, *11*(4), 605 (1972).
114. K. U. Ingold, F. J. Stubbs, and C. N. Hinshelwood, *Proc. R. Soc., London, Ser. A*, *A200*, 458 (1950).
115. K. U. Ingold, F. J. Stubbs, and C. N. Hinshelwood, *Proc. R. Soc., London, Ser. A*, *A201*, 18 (1950).
116. M. Hirato, S. Yoshioka, and M. Tanaka, *Hitachi Rev.*, *20*(8), 325 (1971).
117. M. Hirato and S. Yoshioka, *Int. Chem. Eng.*, *13*(2), 347 (1973).
118. V. Illes, O. Szalai, and Z. Csermely, in *Industrial and Laboratory Pyrolysis* (ACS Symposium Series 32), American Chemical Society, Washington, D.C., 1976, Chap. 24.
119. M. E. Brooks and J. Newman, *Gas Oil Cracking, The Problems That Had To Be Solved*, Presented at VIII. World Petroleum Congress, Moscow, USSR, June 1971.
120. M. E. Brooks and J. Newman, *Oil Gas J.*, *69*(37), 94 (1971).
121. S. B. Zdonik, E. J. Bassler, and L. P. Hallee, *Hydrocarbon Process.*, *53*(2), 73 (1974).
122. L. E. Chambers and W. S. Potter, *Hydrocarbon Process.*, *53*(3), 95 (1974).
123. S. B. Zdonik, G. L. Hayward, S. H. Fishtine, and J. C. Feduska, *Hydrocarbon Process.*, *54*(12), 111 (1975).
124. M. R. Kitzen, F. M. Wall, and H. J. Cijfer, *Chem. Eng. Prog.*, *65*(7), 71 (1969).
125. L. Marshall and H. B. Zasloff, *Chem. Eng. Prog.*, *65*(10), 65 (1969).
126. J. R. Lambrix, C. S. Morris, and H. J. Rosenfeld, *Chem. Eng. Prog.*, *65*(11), 65 (1969).
127. B. P. Ennis, H. B. Boyd, and R. Orriss, *High Temperature, Low Contact Time Pyrolysis* (Institute of Chemical Engineers Symposium Series 43), Harrogate, England, June 1975.

128. Y. M. Zhorou, I. I. Vasileva, G. M. Panchenkov, and S. T. Kuzmin, "Mathematical Description of Naphtha-Pyrolysis Process," translated from *Khim. Tekhnol. Topl. Masel*, *4*, 16 (1974) by Plenum Publishing Corp.
129. E. J. Green, S. B. Zdonik, and L. P. Hallee, *Hydrocarbon Process.*, *54*(9), 164 (1975).
130. S. B. Zdonik, G. L. Hayward, S. H. Fishtine, and J. C. Feduska, *Hydrocarbon Process.*, *55*(1), 154 (1976).
131. H. G. Davis and R. G. Keister, in *Industrial and Laboratory Pyrolysis*, (ACS Symposium Series 32), American Chemical Society, Washington, D.C., 1976, Chap. 22.
132. J. M. Fernandez-Baujin, and S. M. Solomon, in *Industrial and Laboratory Pyrolysis* (ACS Symposium Series 32), American Chemical Society, Washington, D.C., 1976, Chap. 20.
133. S. B. Zdonik, E. J. Green, and L. P. Hallee, "Manufacturing Ethylene," reprinted from the *Oil & Gas Journal* (1966–1970), Petroleum Publishing Co., Tulsa, Oklahoma, 1970, Chap. 5.
134. H. G. Davis and K. D. Williamson, *Proceedings of the Vth World Petroleum Congress, New York, May 31–June 5, 1959*, Vol. IV, 1960, p. 37.
135. K. J. Laidler and B. W. Wojciechowski, *Proc. R. Soc. London, Ser. A.*, *260*, 91 (1961).
136. J. E. Taylor and D. M. Kulich, in *Industrial and Laboratory Pyrolysis* (ACS Symposium Series 32), American Chemical Society, Washington, D.C., 1976, Chap. 5.
137. W. L. Nelson, *Petroleum Refining Engineering*, 4th ed., McGraw-Hill, New York, 1958, Chap. 19.
138. K. D. Williamson and H. G. Davis, in *Industrial and Laboratory Pyrolysis* (ACS Symposium Series 32), American Chemical Society Washington, D.C., 1976, Chap. 4.
139. Y. C. Sun, *Kinetic Equations for Naphtha Cracking*, Presented at AIChE 61st Annual Meetings, Los Angeles, California, December 1–5, 1968.
140. P. S. Van Damme, G. F. Froment, and S. Barendregt, *Thermal Cracking of Naphtha, Kinetics and Product Distributions*, Presented at AIChE 83rd National Meeting, Houston, Texas, March 23, 1977.
141. M. Hirato and S. Yoshioka, *Int. Chem. Eng.*, *13*(2), 347 (1973).
142. S. B. Zdonik, E. J. Green, and L. P. Hallee, "Manufacturing Ethylene," reprinted from the *Oil & Gas Journal* (1966–1970), Petroleum Publishing Co., Tulsa, Oklahoma, 1970, Chap. 6.
143. M. Murata and S. Saito, *J. Chem. Eng.* (*Jpn.*), *8*(1), 39 (1975).
144. H. G. Davis and T. J. Farrell, *Ind. Eng. Chem., Process Des. Dev.*, *12*(2), 171 (1973).
145. H. H. Voge and G. M. Good, *J. Am. Chem. Soc.*, *71*, 593, (1949).
146. G. V. Froment, B. O. Van DeSteene, P. T. Vanden Berghe, and A. G. Goossens, *AIChE J.*, *23*(1), 93 (1977).
147. W. E. Slater, *J. Chem. Soc.*, *109*, 160 (1916).
148. L. S. Lobo and D. L. Trimm, *J. Catal.*, *29*, 15 (1973).
149. J. E. Lanett and G. Egloff, *Ind. Eng. Chem.*, *9*, 350 (1917).
150. J. S. Conroy, R. S. Slysh, D. B. Murphy, and C. R. Kinney, in *Proceedings, Third Conference on Carbon*, Pergamon Press, Oxford, England, 1959, p. 395.
151. R. O. Grisdale, A. C. Pfister, and W. Van Roesbroeck, *Bell System Tech. J.*, *30*, 271 (1959).
152. W. G. Appleby, J. W. Gibson, and G. M. Good, *Ind. Eng. Chem., Process Des. Dev.*, *1*(2), 102 (1962).

153. T. M. John and B. M. Wojciechowski, *J. Catal.*, *37*, 240 (1975).
154. K. I. Syskov, *Jelikhovaskaya*, *Carbon*, *5*, 201 (1967).
155. J. J. Dunkleman and L. F. Albright, in *Industrial and Laboratory Pyrolysis* (ACS Symposium Series 32), American Chemical Society, Washington, D.C., 1976, Chap. 14.
156. J. J. Dunkleman and L. F. Albright, in *Industrial and Laboratory Pyrolysis* (ACS Symposium Series 32), American Chemical Society, Washington, D.C., 1976, Chap. 15.
157. C. H. Tsai and L. F. Albright, in *Industrial and Laboratory Pyrolysis* (ACS Symposium Series 32), American Chemical Society, Washington, D.C., 1976, Chap. 16.
158. S. M. Brown and L. F. Albright, in *Industrial and Laboratory Pyrolysis* (ACS Symposium Series 32), American Chemical Society, Washington, D.C., 1976, Chap. 17.
159. R. E. Kinney and D. J. Crowley; *Ind. Eng. Chem.*, *46*(2), 258 (1954).
160. Y. T. Shah, E. B. Stuart, and K. D. Sheth, *Ind. Eng. Chem.*, *Process Des. Dev.*, *15*(4), 518 (1976).
161. J. L. De Blieck and A. G. Goossens, *Hydrocarbon Process.*, *50*(3), 76 (1971).
162. A. Mol, *Hydrocarbon Process.*, *53*(7), 115 (1974).
163. L. E. Chambers and W. S. Potter, *Hydrocarbon Process.*, *53*(1), 121 (1974).
164. T. Kunugi, H. Tominaga, and S. Abiko, *Pyrolysis of Hydrocarbons in the Presence of Hydrogen*: *Selective Production of Ethylene*, Presented at the VIIth World Petroleum Congress, Mexico City, Mexico, April 1967.
165. E. Chahvekilian, *Olefins by Pyrolytic Cracking of Hydrocarbons under Pressure and in the Presence of Hydrogen*, Presented at AIChE 83rd National Meeting Houston, Texas, March 1977.
166. M. W. Clark, *The Materials Barrier in Pyrolysis Furnace Design and Operation*, Presented at Conference of National Association of Corrosion Engineers, March 4, 1970.
167. S. B. Zdonik, E. J. Green, and L. P. Hallee, "Manufacturing Ethylene," reprinted from the *Oil & Gas Journal* (1966–1970), Petroleum Publishing Co., Tulsa, Oklahoma, 1970.
168. J. L. James, M. E. Hopkins, and R. Orriss, *The Optimization of Pyrolysis Furnace Yields by Temperature Profile Control*, Presented at Journee de'Etude sur le Steam Cracking, Association Francaise des Technicians du Petrole, Paris, France, November 6, 1972.
169. J. M. Fernandez-Baujin and A. J. Gambro, *Technology and Economics for Modern Olefins Plants*, Presented at VIth Interamerican Congress of Chemical Engineers, Caracas, Venezuela, July 13–16, 1975.
170. B. Lohr and H. Dittmann, *Oil Gas J.*, *75*(27), 53 (1977).
171. S. M. Solomon, *Developments in Olefins Production Technology*, Presented at the Institution of Chemical Engineers London Congress, London, England, May 11–12, 1977.
172. Private communication to K. Stork, March 31, 1977.
173. T. Sato and T. Katada, *Chem. Econ. Eng. Rev.*, *4*(7), 14 (1972).
174. T. Sato et al., *Mitsubishi Tech Rev.*, p. 75 (October 1973).
175. H. R. Knuelle, *Chem. Eng. Prog.*, *68*(7), 53 (1972).
176. K. Arisaki, T. Hiruta, S. Kawamura, and M. Yonezawa, *Chem. Econ. Eng. Rev.*, *8*(7/8), 33 (1976).
177. A. J. Gambro, K. Muenz, and M. A. Abrahams, *Hydrocarbon Process.*, *51*(3), 73 (1972).
178. K. Stork, M. A. Abrahams, and A. Rhoe, *Hydrocarbon Process.*, *53*(11), 157

(1974).

179. S. Sinkar, *Oil Gas J.*, *74*(8), 103 (1976).

180. M. C. Sze and N. C. Kafes (C-E Lummus), U.S. Patent 3,720,729 (March 13, 1973).

181. D. T. Davis, T. G. Glover, and J. R. Jones, (B. P. Chemicals International), U.S. Patent 3,781,195 (December 25, 1973).

182. S. B. Zdonik, G. L. Hayward, S. H. Fishtine, and J. C. Feduska, *Hydrocarbon Process.*, *55*(1), 149 (1976).

183. J. J. Leonard, J. E. Gwyn, and G. R. McCullogh, in *Industrial and Laboratory Pyrolysis* (ACS Symposium Series 32), American Chemical Society, Washington, D.C., 1976, Chap. 18.

184. Anonymous, *Oil Gas J.*, *75*(50), 34 (1977).

185. D. Pownall, *Pet. Times*, *81*(2060), 21 (1977).

186. B. Lohr and H. Dittmann, *Oil Gas J.*, *75*(43), 65 (1977).

187. T. Ishikawa and R. G. Keister, *Advanced Cracking Reactor, a Petrochemical Alternative*, Presented at NTRA National Meeting, San Antonio, Texas, April 4, 1978.

188. J. C. Davis, *Chem. Eng.* *85*(1), 25 (1978).

189. A. M. Brownstein, *Chem. Eng. Prog.*, *71*(9), 72 (1975).

190. J. J. McKetta (ed.), *Encyclopedia of Chemical Processing and Design*, Vol. 1, Dekker, New York, 1976 p. 114.

191. S. A. Miller, *Ethylene and Its Industrial Derivatives*, Baum, London, 1969, p. 864.

192. F. Applegath et al., (Monsanto), U.S. Patent 3,848,012 (November 12, 1974).

193. A. C. MacFarlane, *Oil Gas J.*, *74*(6), 99 (1976).

194. F. G. Dwyer, P. J. Lewis, and F. H. Schneider, *Chem. Eng.* (*N.Y.*), *83*(1), 90 (1976).

195. P. J. Lewis and F. G. Dwyer, *Oil Gas J.*, *75*(40), 55 (1977).

196. W. Tucker and W. E. Cline, *Plant Reliability and Its Impact on Large Plant Design and Economics*, Presented at Third Joint Meeting of Instituto de Ingenieros Quimicos de Puerto Rico and AIChE, San Juan, Puerto Rico, May 1970.

197. R. W. Henry, *Let's Profit from Experience*, Presented at AIChE 84th National Meeting, Atlanta, Georgia, February 26–March 1, 1978.

198. J. L. Boyett, *Finding and Defining Booby Traps Built into Ethylene Plants*, Presented at AIChE 84th National Meeting, Atlanta, Georgia, February 26–March 1, 1978.

199. J. Barnwell, *Inherent Safety Design Features for an Ethylene Plant*, Presented at AIChE 84th National Meeting, Atlanta, Georgia, February 26–March 1, 1978.

200. T. A. Kletz, *Practical Applications of Hazard Analysis*, Presented at AIChE 84th National Meeting, Atlanta, Georgia, February 26–March 1, 1978.

201. Anonymous, *Pet. Times*, *82*(2070), 10 (March 3, 1978).

202. J. J. McKetta (ed.), *Encyclopedia of Chemical Processing and Design*, Dekker, New York.

203. J. J. McKetta (ed.), *Encyclopedia of Chemical Processing and Design*, Vol. 3, Dekker, New York, 1977, p. 1.

204. H. Benzler and A. v. Koch, *Chem.-Ing.-Tech.*, *27*(2), 71 (1955).

205. K. M. Sundaram and G. F. Froment, *Ind. Eng. Chem., Fundam.*, *17*(3), 174 (1978).

206. L. F. Albright, *Ind. Eng. Chem., Process Des. Dev.*, *17*(3), 377 (1978).

207. A. G. Goossens, M. Dente, and E. Ranzi, *Oil Gas J.*, *76*(36), 89 (1978).

208. L. F. Albright and C. F. McConnell, *Deposition and Gasification of Coke during Ethane Pyrolysis*, Presented at ACS 175th National Meeting, Anaheim, California, March 12–17, 1978.

209. L. F. Albright, C. F. McConnell, and K. Welther, *Types of Coke Formed during the Pyrolysis of Light Hydrocarbons*, Presented at ACS 175th National Meeting, Anaheim, California, March 12–17, 1978.

210. L. F. Albright and C. Y.-H. Yu, *Pyrolysis of Acetylene, Butadiene and Benzene in Different Tubular Reactors*, Presented at ACS 175th National Meeting, Anaheim, California, March 12–17, 1978.

211. O. Winter, *Ethylene from Ethanol*, Presented at First Brazilian Petrochemical Congress, Rio de Janeiro, November 7–12, 1976.

212. A. M. Brownstein, *Trends in Petrochemical Technology*, Petroleum Publishing Co., Tulsa, Oklahoma, 1976, pp. 127 and 69–76.

213. B. Buessemeier, C. D. Frohning, and B. Cornils, *Hydrocarbon Process.*, *55*(11), 105 (1976).

214. W. W. Kaeding and S. A. Butter, (Mobil Oil Corp.), U.S. Patent 3,911,041 (October 7, 1975).

215. J. E. Phillips (Phillips Petroleum Co.), U.S. Patent 3,236,912 (February 22, 1966).

216. Private communication to K. Stork, May 19, 1977.

217. B. Alberts and D. P. Zwartbol, *Oil Gas J.*, *77*(23), 137 (1979).

Index

F

Q

R

S

T

V

W

X

Y